ENCYCLOPAEDIA OF

PLANT ECOLOGY

Encyclopaedia of
Plant Ecology

Volume 3

A.Z. Bukhari

ANMOL PUBLICATIONS PVT. LTD.
NEW DELHI - 110 002 (INDIA)

ANMOL PUBLICATIONS PVT. LTD.
H.O.: 4374/4B, Ansari Road, Darya Ganj,
New Delhi-110 002 (India)
Ph.: 23278000, 23261597
B.O.: No. 1015, Ist Main Road, BSK IIIrd Stage
IIIrd Phase, IIIrd Block
Bangalore - 560 085 (India)
Visit us at: www.anmolpublications.com

Encyclopaedia of Plant Ecology

First Published, 2007
ISBN 81-261-3080-6 (Set)

PRINTED IN INDIA

Printed at Mehra Offset Press, Delhi.

Contents

Preface

Ecology is the study of relations between plants and living creatures and the surrounding environment. This relation is vital in itself and indicates how significant and necessary plants and animals are for each other. Biologically, each fulfills another's demand for life. This is a multi-faceted issue.

Animals' respiration is based on two types of gases: Oxygen and Carbon dioxide. When we breath in, we take Oxygen inside, released by plants. On the contrary, when we breathe out, we release Carbon dioxide, which is absorbed by plants. In other words, no oxygen, no animal life; no carbon dioxide, no plants.

Plants also serve as an important source of energy for animals in the form of food or fuel. Similarly, animal's excreta is used in fusing new vitality into soil to help seeds grow into plants and trees. Thus, both constitute a never-ending cycle. Besides, there are various other elements, which play important roles in creating a friendly environment for animals, as well as plants.

In fact, it is a vast knowledge which is wellnigh impossible to discuss here. A detailed and research-based study of the subject has been attempted in the form of this multi-volume work, namely, *Encyclopaedia of Plant Ecology*. Researchers students, scholars, academics and the general readers would expectedly find it a useful tool of information in their possession.

We await and admire comments and remarks of the intelligent readers.

— Editor

1

Role of Earth

A. How Universe came into being?

"Big bang", 8×10^9 years ago, estimated from galaxy expansion and background temperature of microwaves. Recent hypothesis suggests "bubble" inflating into different dimensions from a discontinuity ("white hole"-a tiny region of ultra high density and temperature) in a previous universe.

B. How Solar System came into being?

Aggregation of cold gas and dust by gravity $4\text{-}5 \times 10^9$ years ago, estimated by radioactive decay (ratios of radioactive elements to their decomposition products).

Original gas (stellar): H_2, He, NH_3, H_2O, CH_4

Spinning of matter caused gravitational attraction—condensation of matter, increases spinning and gravitational attraction thereby attracting more matter—eventually, sun and planets formed via accretion of materials. Small planets lose H_2, He; gravitation and radioactive decay heats their cores, expelling N_2 (NH_3), H_2O, CO_2, CO, (CH_4) into the atmosphere; also H_2S, H_3PO_4 and rocks.

Inner Planets—heavier elements with high melting and boiling temperatures Earth-Composition- MgO, SiO_2, FeO, Fe, Ni, some water, NH_4

Outer Planets - lighter elements with low melting and boiling temperatures

Jupiter, Saturn, Uranus, Neptune - mainly composed of hydrogen, methane, ammonia ices Earth and Oceans formed 4.5 billion years ago

- decay of radioactive elements generates lots of heat
- gravity pulled most of the iron inward to form the planets core
- lighter elements, silicon, magnesium, aluminium, and oxygen bonded to form less dense compounds that migrated toward the surface
- by 4.0 billion years ago earth's surface cooled enough for water to exist primarily as a liquid. Atmosphere formed - formed 4.5 to 3.5 billion years ago - primary constituents of the atmosphere: methane, ammonia, carbon dioxide, water vapor. Carbon dioxide combines with water to form carbonic acid

  $H_2O + CO_2\ H_2CO_3$

which dissolves rocks releasing minerals (nutrients), or the essential building blocks needed for life.

C. Organic Molecules Creation

Simple chemical reactions, requiring only raw materials and energy were the first step in the origin of life on earth.

Origin of Life: It took place 3.5 billion years ago.

Definition of life - capable of growing more complex, obtains and use energy by breaking down chemical compounds, capable of self replicating. Water is an ideal medium for biochemical reactions, its unique properties make it a prerequisite for origin of life, some of them are:

1. It retains heat
2. Very good solvent - dissolves many chemicals
3. Very good transport medium for nutrients

Primordial soup was created from a mixture of water vapor, ammonia, methane, and hydrogen.

Possible sources of energy for the chemical reactions:

1. Electrical discharge
2. Ultraviolet light
3. Geothermal and hydrothermal heat (ocean floor)

D. How Self-producing System came into being?

Amino acids, nucleotides form polymers: some theoreticians suggest that life must have both self-reproducing templates (nucleic acids) and catalysts (proteins) -which came first? How did the two get together?

Note that some RNAs (viral) are self-templates and some show enzymatic activity (RNA as selfsplicing). Because of this, some speculate that RNA was the first self-reproducing system. Some amino acids, trinucleotides (model colons) show affinities, suggesting a natural system leading to nucleic acid-directed protein synthesis. DNA evolved as a more stable form of template.

E. Gradual Transformation of Energy Metabolism

Strategy: use easiest energy sources first.

1. Fermentation (*e.g.*, anaerobic bacteria)

   Energy from UV, lightning stored as covalent bonds in soup components. First cells obtained energy by breaking down existing molecules (non-oxidative, because there was no free O_2). Eventually, cells would run out of substrate, since life grows faster than synthesis works.

2. Photosynthesis (*e.g.*, blue-green algae)

   Use visible light (not UV) as energy source (visible penetrates through H_2O farther than UV). Use H_2, H_2S and eventually H_2O as source of electrons and H+, depending on abundance. When HO used, O_2 produced; production of O_2 leads to formation of O_3, shielding of UV opening of new niches in shallow water, land.

3. Aerobic respiration (*e.g.*, aerobic bacteria)

   Use energy stored in organic compounds. O_2 released by the oxidative reaction. Use is made possible by H_2O photosynthesis; possibly started in plants as protection against O_2. Efficient aerobic respiration allowed evolution of consumers.

F. Transformation of Complex Organelles (eucaryotes)

1. *Endosymbiotic Hypothesis:* Some organelles like plastids, mitochondria have double membranes, their own DNA, their own ribosomes, bacterial-like protein synthesis (similar but not identical). Suggest plastids formed by phagocytosis of photosynthetic procaryote, with stabilized symbiotic relationship (food, protection). Similarly mitochondria formed the same way with O_2-using procaryote.

2. *Packaging of Pre-existing Cell Components Hypothesis:* Photosynthetic cell with genes for chlorophyll, photosynthetic electron transport forms nucleus by invagination of plasma membrane; nucleus buds off some DNA to make accessory nucleus with thylakoids; mitochondria are evolved as modified chloroplasts.

Earth: For study purposes earth can be divide into following groups:

Geosphere: The geosphere is the solid Earth that includes continental and oceanic crust as well as the various layers of the Earth's interior. Ninety-four per cent of the earth is composed of the elements oxygen, iron, silica, and magnesium, and the interior of the earth is layered both chemically and mechanically. The geosphere is not static, but its surface (crust) is in a constant state of motion that gives rise to movement of the continents. The unifying theory that explains how continental drift is possible is called plate tectonics, it was developed by geologists in the 1960s.

Atmosphere: Each year volcanism spews significant amounts of gases (water vapor, carbon dioxide, sulphur dioxide, hydrogen, hydrogen sulphide, etc.) into the atmosphere. For example, explosive volcanism that injects large amounts of sulphur dioxide to the upper atmosphere is known to result in cooling of climate in the years immediate following the eruption.

Hydrosphere: Water is the main agent of chemical and mechanical erosion of the earth surface. When water flows over land it mechanically breaks rocks into finer particles and can chemically dissolve elements contained in rock-forming minerals. The sediment and chemical constituents are carried away downstream as the water makes its way back to the ocean. For example, the awesome depth (in Kilometers) of the Grand Canyon which was cut by the erosive power of the Colorado River. Over long periods of geologic time, erosion by water can wear down mountain ranges (process called peneplanation) and transport this material to be deposited in sedimentary basins.

Biosphere: The weathering of the geosphere to form soils provides terrestrial plants with a firm substrate and vital nutrients (phosphorous, nitrogen) and minerals needed for plant growth. In addition, the chemical weathering of the geosphere by water transports essential nutrients (phosphorous, nitrogen, silica, etc.) to the oceans which are used by algae (marine plants) during photosynthesis.

Various Ecology Systems

An ecosystem consists of all of the organisms living in a community plus all of the abiotic factors with which they interact. As such, ecosystems are the most important level of biological organization.

Trophic Structure: Trophic structures determine the feeding relationships between organisms and the energy and chemical cycling pathways. Trophic structure is typically divided into several trophic levels. These levels are arranged hierarchically. The trophic level that supports all others are the primary producers of the ecosystem. These are the photosynthetic autotrophs that use sunlight to create complex carbohydrates.

All other organisms in an ecosystem are consumers, or heterotrophs, that depend directly or indirectly on the primary producers. Those that depend directly on primary producers are the organisms in the next trophic level, the primary consumers (herbivores). Animals that feed on herbivores (carnivores) occupy the next higher trophic level of the secondary consumers. Animals that feed on secondary consumers are tertiary consumers, and so on. Some consumers feed on the waste products or dead remains of members of other trophic levels. These are the decomposers (detritivores). The simplest arrangement of these trophic levels is: Primary producer - primary consumer - secondary consumer - tertiary consumer this is a form of a food chain. More often than not there are many organisms interacting at different trophic levels so interactions in the real world more often resemble a food web.

I. Energy in Ecosystems

1. *Global Energy Budget:* The primary source of energy for nearly all ecosystems on earth is the sun. Only a fraction

of the sun's energy that strikes the earth reaches photosynthetic organisms and of this amount, about 1-2% is converted to chemical energy by photosynthesis. The amount of sunlight converted to chemical energy by primary producers during any given time is known as primary productivity.

Different ecosystems have different productivities. Tropical rain forests and algal beds and reefs are extremely productive ecosystems.

2. *Limits to Productivity:* Seasonal limits like rainfall, temperature, and light intensity are very important. Inorganic nutrients are also important. If a single factor is limiting it is known as the limiting nutrient. Replacing limiting nutrients in agricultural systems is at the heart of the notion of fertilizing.

3. *Energy Loss:* Energy is lost as it flows from one trophic level to the next. No consumer is ever 100% efficient in converting resources obtained from one trophic level and converting it into the energy needed to sustain life. There is only about a 10% transfer of energy from one trophic level to the next. For example, a caterpillar might consume 200 Joules of energy in plant material. Of that 200 J, 100 J might be eliminated as feces, 67 J as respiration, and just 33 J used for growth. It is that 33 J used for growth that will be available to an organism on another trophic level that eats the caterpillar. The differential between respiration and growth is even more extreme in homeotherms (like humans) that use a great deal of the energy they consume to maintain body temperature.

Ecological efficiencies (the fraction of energy transferred from one trophic level to the next) can be thought of in terms of a pyramid of productivity, where the trophic levels are stacked in blocks. Pyramids like this are bottom heavy, with either much more biomass or many more numbers of organisms in the lower levels than the upper ones. This arrangement reflects the inherent inefficiencies that exist in transferring energy from one level to the next.

II. Biogeochemical Cycles

Life on earth receives the energy it needs from the sun which is, for all practical purposes, inexhaustible. The other requirements of life, in other words the essential chemical elements, are present in fixed amounts and so need to be constantly recycled. Chemicals that undergo biogeochemical cycles include water, nitrogen, phosphorus, carbon, etc.

A general model of biogeochemical cycling includes four parts divided between inorganic and organic materials and whether or not the chemical is available or unavailable for use by living things. These four parts, or compartments, are thus:

- Organic available: living things and detritus.
- Organic unavailable: typically fossilized living things (oil, peat, coal, etc.).
- Inorganic available: chemicals dissolved in water or present in the air that organisms can use.
- Inorganic unavailable: chemicals tied up in rocks.

Nutrient cycling is therefore the relative movement of matter between and among the various compartments.

A. *The Water Cycle:* All living things need water. Water also participates in all of the other chemical cycles. Most of the earth's water is present in the oceans. Most of it evaporates from and precipitates over these same oceans. Some evaporated water reaches land and falls as rain. Here some evaporates again. A larger fraction is turned into water vapor by plant transpiration. Another fraction of it percolates into the soil or becomes involved in groundwater and runoff. Eventually, this water reaches the oceans where rivers empty into them.

B. *The Carbon Cycle:* Carbon is found in all organic compounds. The basic carbon cycle involves the assimilation of carbon (as CO_2) by plants during photosynthesis and the return of carbon to the atmosphere by respiration and decomposition (again, as CO_2). There is a seasonal fluctuation in the carbon cycle due to winter in the Northern Hemisphere (much less carbon is fixed in the winter time). Carbon also flows into aquatic systems by combining with water to form carbonic acid (H_2CO_3) and in turn, bicarbonate ion. The amount of carbon present in aquatic ecosystems (essentially the oceans) is about 50 times as great as the amount present in the atmosphere. Some carbon is removed from the carbon cycle when dead organisms are turned into coal, oil, or peat. Humans are affecting the global carbon cycle by burning large amounts of otherwise unavailable carbon when we use fossil fuels. The CO_2 released in this way is increasing the CO_2 concentration in the atmosphere.

C. *The Nitrogen Cycle:* Nitrogen is a fundamental component of proteins. Most nitrogen is directly assimilated from the environment by plants. Nitrogen (as N_2) makes up about 80% of the earth's atmosphere but this gas is an unavailable source of nitrogen for plants. To obtain nitrogen in a usable form plants need nitrate (NO_3^-). They obtain this in two ways. A family of plants called Legumes has established a mutualistic relationship with nitrogen-fixing bacteria that lets them obtain nitrogen directly from atmospheric nitrogen. Most other plants obtain (NO_3^-) via the nitrification of ammonium ion, which itself comes from decomposition by fungi and bacteria (through ammonification) and the direct fixation of nitrogen gas by some bacteria.

Consumers obtain nitrogen from producers and other consumers depending on the trophic level. Nitrogen gas' is returned to the atmosphere by denitrifying bacteria.

Although nitrogen cycles between the atmosphere and living things, most nitrogen is recycled locally through assimilation and decomposition.

III. Human Influence on Ecosystems

A. *Poisons in Food Chains:* Humans create a lot of different chemical compounds that are released intentionally or by accident into the environment. Some of these are toxic. In other cases, action by microorganisms in the environment create toxic compounds from them. Some of these compounds can persist for years in the environment and end up being stored in the tissues of the organisms that eat them.

For example, two classes of compounds containing carbon-chlorine bonds are particularly pernicious. Organochlorine insecticides (*e.g.*, DDT) and PCBs (polychlorinated biphenyls) were produced in vast quantities after World War II. Both these compounds are fat-soluble and so they have accumulated in the fatty tissue of the organisms that have ingested them. Animals that eat the organisms that contain these compounds can themselves be contaminated, often with very high doses. The increasing concentrations of organochlorines and PCBs that are found in animals that are at ever higher levels in the food chain is known as biological magnification. For example, DDT might be present in ca. 0.000003 ppm (parts per million) in a given water sample but might be as high as 25 ppm by the time it reaches the top carnivore in a food chain. Humans are not immune to this phenomenon. There is probably no one on earth today who is not carrying at least some of these chemicals in their fatty tissue. These chemicals can even be passed on to newborns via the milk of nursing mothers.

B. Carbon Dioxide Emissions and the Greenhouse Effect-an alteration of the carbon cycle. Modern industrial society is based largely on the burning of fossil fuels for energy. Fossil fuels are sources of hydrocarbons that were formed from the decayed remains of organisms millions of years ago. They include oil and coal. When these are burned for their energy CO_2 is released as a by-product. The level of CO_2 in the atmosphere has gradually increased over time due, apparently, to this release of CO_2 by human activity. This increasing CO_2 concentration has been correlated with an increase in average global temperature. CO_2 is responsible for retaining heat in our atmosphere. The more of it there is, the warmer the earth will be. If the average temperature of the earth increases enough (and his need only be ca. 3-4 degrees C), large-scale melting of the polar ice caps could result with the consequent flooding of many low lying areas, including many of world's most populous cities. The earth is undeniably warming, but scientists are still uncertain about the degree to which this is human influenced.

C. *Depletion of the Ozone Layer:* One product of free oxygen in the atmosphere is ozone (O_3). Ozone is harmful in high concentrations when we breathe it, but high in the atmosphere it serves a protective role by absorbing UV light, which is very energetic and harmful to living things. The amount of ozone in the atmosphere is decreasing, probably as the result of reactions with human made chemicals called chlorofluorocarbons. These are chemicals that had until recently been used as refrigerants and propellants in aerosol cans. These chemicals can react with ozone to destroy it.

"Ozone holes" have been documented at high latitudes and these grow and shrink seasonally. There are now production bans on these chemicals in many parts of the world. A thinning of the ozone would mean more ultraviolet light exposure for living things and a potential increase in mutation levels.

D. *Human-caused Extinctions:* Extinctions are a normal part of life on planet earth but seldom has the rate of extinction been as great as it is now, and it is primarily due to human activity. As human population increases so does its demand on the world's resources. One of these resources is space and humans are encroaching on the natural world to a greater degree each year. That encroachment means habitat loss for many species and as habitat is lost, many species go extinct. In some places, like the rain forests, species are being lost so quickly that the rate might be as high as 1000-10,000 times what it normally is.

Biodiversity Hotspots: Biodiversity is not evenly distributed on the planet. Some areas harbor far greater concentrations of living creatures than others. Scientists had identified those biologically rich areas under the greatest threat of destruction as biodiversity hotspots, and are concentrating conservation in these areas to minimize it. Approximately 25 priority hotspots have been identified based on three criteria: the number of species present, the number of those species found exclusively in an ecosystem and the degree of threat they face.

History of the Hotspots: British ecologist Norman Myers created the biodiversity hotspots concept in two scientific papers published in 1.988 and 1990. Myers recognized that a modest number of hotspot ecosystems covering a small total land area, most often in tropical forest areas, accounted for a high percentage of global biodiversity.

Areas of the Hotspots

1. Tropical Andes
2. Mediterranean Basin
3. Madagascar/Indian Ocean Islands
4. Mesoamerica
5. Caribbean Islands
6. Indo-Burma
7. Atlantic Forest of Brazil
8. Philippines
9. Cape Floristic Region of South Africa
10. Mountains of South Central China
11. Sundaland
12. Brazilian Cerrado
13. Southwest Australia
14. Polynesia and Micronesia
15. New Caledonia
16. Choco/Darien/Western Ecuador
17. Western Ghats and Sri Lanka
18. California Floristic Province
19. Succulent Karoo
20. New Zealand
21. Central Chile
22. Guinean Forests of West Africa
23. Caucasus
24. Eastern Arc Mountains, Coastal Forests of Kenya and Tanzania
25. Wallacea

Some facts about biodiversity:

- Diversity increases from polar zones to tropics which is true for almost all taxa.
- Most species found on the earth are in tropical rainforests.
- Tropical rain forests cover 6% of land area and contain 50% of all species.
- Diversity is high in the tropics due to favourable Productivity, Habitat heterogeneity, Niche specialization, Population interactions, Climate history.
- About 60 million years ago - most of Earth was tropical forest. Since Cenozoic era there has been a continuous cooling and drying of global climate which led to evolution of new biomes - temperate forests, grasslands, deserts, etc.
- Lineages in tropical forests have been evolving for more than 60 million years- while lineages that are adapted to cooler and drier conditions are younger.
- Ice ages in northern latitudes during last 2 million years depleted diversity available at that time.
- Each biome within each biogeographic realm has endemic species - species found nowhere else.
- Now humans have modified 60% of land surface.
- Habitat fragmentation leads to extinction.
- 99.9% of all species found at some time on earth have gone extinct.
- The background extinction rate of species is 1/million species/year.
- Given a current species diversity of 1 - 10 million, we should see 1-10 extinctions per year. But the current extinction rate is much higher - *e.g.*, birds. Of 10,000

species of birds (1-10% of total number of species) approximately 2,000 became extinct in last 2,000 years. There is good evidence that birds are "typical" in their risk of extinction—there are as more extinctions in other taxa.

- It is estimated that the current rate of extinction is around 17,000 - 100,000 extinctions per year. An average estimate is 35,000/year which is similar to past mass extinctions.
- Leading cause of extinction is tropical deforestation.
- Rate of tropical deforestation is 1-3%. At this rate 50% of all species present on earth may be lost by 2050.

Successful action to conserve biodiversity must address the full range of causes of its current loss and embrace the opportunities that genes, species, and ecosystems provide for sustainable development. Because the goal of biodiversity conservation-supporting sustainable development by protection and using biological resources in ways that do not diminish the world's variety of genes and species or destroy important habitats and ecosystems is so broad, any biodiversity conservation strategy must also have a broad scope. A biodiversity campaign can be broken down into three basic elements:

- saving biodiversity,
- studying it, and
- using it sustainably and equitably.

A. *Saving Biodiversity:* Saving biodiversity means taking steps to protect genes, species, habitats, and ecosystems. The best way to maintain species is to maintain their habitats. Saving biodiversity therefore often involves efforts to prevent the degradation of key natural ecosystems and to manage and protect them effectively. But since many of the world's habitats have been modified for such human uses as agriculture, the programme must include measures to maintain diversity on lands and in waters that have already been disturbed. A third component is restoring lost species to their former habitats and preserving species in gene-banks, zoos, botanic gardens, and other off-site (ex situ) facilities.

B. *Studying Biodiversity:* Studying biodiversity means documenting its composition, distribution, structure, and function; understanding the roles and functions of genes, species, and ecosystems; grasping the complex links between modified and natural systems; and using this understanding to support sustainable development. It also means building awareness of biodiversity's values, providing opportunities for people to appreciate nature's variety, integrating biodiversity issues into educational curricula, and ensuring that the public has access to information on biodiversity, especially on developments that will influence it locally.

C. *Using Biodiversity Sustainably and Equitably:* Using biodiversity sustainably and equitably means husbanding biological resources so that they last indefinitely, making sure that biodiversity is used to improve the human condition, and seeing that these resources are shared equitably. "Use" does not, however, automatically imply consumption. Often, the best economic use of biodiversity may be to maintain it in its natural state for its ecological or cultural values, as in the cases of forested watersheds or sacred groves.

Biodiversity Conservation Scheme: The biodiversity conservation agenda must encompass much more than concern for protected areas, threatened species, zoos or seed-banks and its constituency must be broad-based. It has to take place within the wide context of the move toward sustainable living discussed in *Our Common Future* the report of the World Commission on Environment and Development-and detailed in *Caring for the Earth,* the successor and complement to the *World Conservation Strategy.*

How can biodiversity conservation be addressed within the context of sustainable development, as it must succeed? There must be new contacts and partnerships within communities, bringing biologists and resource managers together with social scientists, political leaders, businessmen, religious leaders, farmers, journalists, artists, planners, teachers, and lawyers. There must be dialogue between central and local governments, industry, and citizen's groups, including non-governmental environment and development organizations, and women's and indigenous people's organizations. New mechanisms for discussion, negotiation, and common action are all essential.

Biodiversity conservation must take place at the individual level, the global level, and in between. Effective conservation efforts begin in the fields, forests, watersheds, grasslands, coastal zones, and settlements where people live and work. But complementary governmental efforts are needed to address the many facets of biodiversity conservation beyond the capacity of local communities, or involving resources that are of national importance. By the same token, international cooperation is essential, given the global nature of the biodiversity crisis and the lack of national resources in many countries.

Many essential elements of biodiversity conservation require sustained commitment, but will not show immediate results. Policies, institutions, laws, and attitudes do not change overnight. Expanding human capacity, carrying out first-rate research, and conducting biodiversity inventories take time and money and may have no immediate pay-off. But they create the larger context in which enduring change can take hold and emergency measures have at least a hope of succeeding. Still, immediate action is needed. Irreplaceable genes, species, and ecosystems are disappearing at a rate unprecedented in human history, and essential development is at risk as a result. Immediate action is needed

- to defend these threatened living resources;
- to reform the policies that invite such losses;
- to conduct inventory and study of resource use in key ecosystems and countries;
- to monitor changes and impeding threats;
- to better manage threaten protected areas;
- to mobilize funding; and
- to support national and grassroot's conservation initiatives.

Making a Biodiversity Strategy: The limited conservation resources available must be focussed strategically on opportunities likely to yield the greatest conservation benefits, and five key strategic objectives offer significant possibilities for effective action:

1. To develop national and international policy frameworks that foster the sustainable use of biological resources and the maintenance of biodiversity;
2. To create conditions and incentives for effective conservation by local communities;
3. To strengthen and apply more broadly the tools for conserving biodiversity;
4. To greatly strengthen the human capacity for conserving and using biodiversity sustainably, particularly in developing countries;

5. To catalyze conservation action through international cooperation and national planning.

Promoting Actions for Biodiversity Conservation

- Ratify the Convention on Biological Diversity like other international treaties.
- Establish a mechanism such as an International Panel on Biodiversity Conservation (preferably within the Convention on Biological Diversity), including scientists, non-governmental organizations, and policy-makers to provide guidance on priorities for the protection, understanding, and sustainable and equitable use of biodiversity.
- Establish an Early Warning Network, linked to the Convention on Biological Diversity, to monitor potential threats to biodiversity and mobilize action against these threats.
- Integrate biodiversity conservation in national planning processes.

Hydro-electric Power: Hydro-electric power plants convert the kinetic energy contained in falling water into electricity. The energy in flowing water is ultimately derived from the sun, and is therefore constantly being renewed. Energy contained in sunlight evaporates water from the oceans and deposits it on land in the form of rain. Differences in land elevation result in rainfall runoff, and allow some of the original solar energy to be captured as hydro-electric power. Hydro power is currently the world's largest renewable source of electricity, accounting for 6% of worldwide energy supply or about 15% of the world's electricity. Traditionally thought of as a cheap and clean source of electricity, most large hydro-electric schemes being planned today are coming up against a great deal of opposition from environmental groups and local people.

The first recorded use of water power was a clock, built around 250 BC. Since that time, humans have used falling water to provide power for grain and saw mills, as well as a host of other applications. The first use of moving water to produce electricity was a waterwheel on the Fox river in Wisconsin in 1882, two years after Thomas Edison unveiled the incandescent light bulb. Early hydro-electric power plants were much more reliable and efficient than the fossil fuel fired plants of the day. This resulted in a proliferation of small to medium-sized hydro-electric generating stations distributed wherever there was an adequate supply of moving water and a need for electricity.

As electricity demand soared in the middle years of this century, and the efficiency of coal and oil-fueled power plants increased, small hydro plants fell out of favour. Most new hydro-electric development is focused on huge "mega-projects".

The majority of these power plants involved large dams which flooded vast areas of land to provide water storage and therefore a constant supply of electricity. In recent years, the environmental impacts of such large hydro projects are being identified as a cause for concern. It is becoming increasingly difficult for developers to build new dams because of opposition from environmentalists and people living on the land to be flooded. This is shown by the opposition to projects such as Tehri and Narmada projects.

Hydro-electric Power Plants: Hydro-electric power plants capture the energy released by water falling through a vertical distance, and transform this energy into useful electricity. In general, falling water is channeled through a turbine which converts the water's energy into mechanical power. The rotation of the water turbines is transferred to a generator which produces electricity. The amount of electricity which can be generated at a hydro-electric plant is dependent upon two factors.

(1) the vertical distance through which the water falls, called the "head", and

(2) the flow rate, measured as volume per unit time.

The electricity produced is proportional to the product of the head and the rate of flow. The following is an equation which may be used to roughly determine the amount of electricity which can be generated by a potential hydro-electric power site:

$$\text{POWER (kW)} = 5.9 \times \text{FLOW} \times \text{HEAD}$$

In this equation, FLOW is measured in cubic meters per second and HEAD is measured in meters. Based on the facts presented above, hydro-electric power plants can generally be divided into two categories. "High head" power plants are the most common and generally utilize a dam to store water at an increased elevation.

The use of a dam to impound water also provides the capability of storing water during rainy periods and releasing it during dry periods. This results in the consistent and reliable production of electricity, able to meet demand. Heads for this type of power plant may be greater than 1000 feet. Most large hydro-electric facilities are of the high head variety. High head plants with storage are very valuable to electric utilities because they can be quickly adjusted to meet the electrical demand on a distribution system.

"Low head" hydro-electric plants are power plants which generally utilize heads of only a few meters or less. Power plants of this type may utilize a low dam or weir to channel water, or no dam and simply use the "run of the river". Run of the river-generating stations cannot store water, thus their electric output varies with seasonal flows of water in a river. A large volume of water must pass through a low head hydro plant's turbines in order to produce a useful amount of power. Hydro-electric facilities with a capacity of less than about

25 MW (1 MW = 1,000,000 Watts) are generally referred to as "small hydro", although hydro-electric technology is basically the same regardless of generating capacity.

"Pumped Storage" is another form of hydro-electric power. Pumped storage facilities use excess electrical system capacity, generally available at night, to pump water from one reservoir to another reservoir at a higher elevation. During periods of peak electrical demand, water from the higher reservoir is released through turbines to the lower reservoir, and electricity is produced. Although pumped storage sites are not net producers of electricity—it actually takes more electricity to pump the water up than is recovered when it is released—they are a valuable addition to electricity supply systems. Their value is in their ability to store electricity for use at a later time when peak demands are occurring. Storage is even more valuable if intermittent sources of electricity such as solar or wind are hooked into a system.

Impact Environmental by Hydro-electric Power Plants: Hydro-electric power plants have many environmental impacts, some of which are just beginning to be understood. These impacts, however, must be weighed against the environmental impacts of alternative sources of electricity. Until recently there was an almost universal belief that hydro power was a clean and environmentally safe method of producing electricity. Hydro-electric power plants do not emit any of the standard atmospheric pollutants such as carbon dioxide or sulphur dioxide given off by fossil fuel fired power plants. In this respect, hydro power is better than burning coal, oil or natural gas to produce electricity, as it does not contribute to global warming or acid rain. Similarly, hydro-electric power plants do not result in the risks of radioactive contamination associated with nuclear power plants.

A few recent studies of large reservoirs created behind hydro dams have suggested that decaying vegetation, submerged by flooding, may give off quantities of greenhouse gases equivalent to those from other sources of electricity. If this turns out to be true, hydro-electric facilities that flood large areas of land might be significant contributors to global warming. Run of the river hydro- plants without dams and reservoirs would not be a source of these greenhouse gases.

The most obvious impact of hydro-electric dams is the flooding of vast areas of land, much of it previously forested or used for agriculture. The size of reservoirs created can extremely be large. Reservoirs can be used for ensuring adequate water supplies, providing irrigation, and recreation; but in several cases they have flooded the homelands of native peoples, whose way of life has then been destroyed. Many rare ecosystems are also threatened by hydro-electric development.

Large dams and reservoirs can have other impacts on a watershed. Damming a river can alter the amount and quality of water in the river downstream of the dam, as well as preventing fish from migrating upstream to spawn. These impacts can be reduced by requiring minimum flows downstream of a dam, and by creating fish ladders which allow fish to move upstream past the dam. Silt, normally carried downstream to the lower reaches of a river, is trapped by a dam and deposited on the bed of the reservoir. This silt can slowly fill up a reservoir, decreasing the amount of water which can be stored and used for electrical generation.

The river downstream of the dam is also deprived of silt, which fertilizes the river's flood plain during high water periods. Bacteria present in decaying vegetation can also change mercury, present in rocks underlying a reservoir, into a form, which is soluble in water. The mercury accumulates in the bodies of fish and poses a health hazard to those who depend on these fish for food. The water quality of many reservoirs also poses a health hazard due to new forms of bacteria, which grow in many of the hydro rivers. Therefore, run of the river type hydro plants generally have a smaller impact on the environment.

Role of Energy

The theoretical size of the worldwide hydropower is about four times greater than that which has been exploited at this time. The actual amount of electricity,' which will ever be generated by hydropower, will be much less than the theoretical potential. This is due to the environmental concerns outlined above and economic constraints. Much of the remaining hydro potential in the world exists in the developing countries of Africa and Asia.

Harnessing this resource would require billions of dollars, because hydroelectric facilities generally have very high construction costs. In the past, the World Bank has spent billions of foreign aid dollars on huge hydroelectric projects in the third world. Opposition to hydro power from environmentalists and native people, as well as new environmental assessments at the World Bank will restrict the amount of money spent on hydroelectric power construction in the developing countries of the world.

In North America and Europe, a large percentage of hydropower potential has already been developed. Public opposition to large hydro schemes will probably result in very little new development of big dams and reservoirs. Small scale and low-head hydro capacity will probably increase in the future as research on low-head turbines, and standardized turbine production, lowers the costs of hydroelectric power at sites with low-heads.

New computerized control systems and improved turbines may allow more electricity to be generated from existing facilities in the future. As well, many small hydro-electric sites were abandoned in the 1950's and 60's when the price of oil and coal was very low, and their environmental impacts unrealized. Increased fuel prices in the future could result in these facilities being refurbished.

Hydro-electric power has always been an important part of the world's electricity supply, providing reliable, cost-effective electricity, and will continue to do so in the future. Hydro power has environmental impacts which are very different from those of fossil fuel power plants.

The actual effects of dams and reservoirs on various ecosystems are only now becoming understood. The future of hydro-electric power will depend upon future demand for electricity, as well as how societies value the environmental impacts of hydro-electric power compared to the impacts of other sources of electricity.

Comparing Energy's Alternative Sources

Solar:

Advantages: Always available, no pollution

Disadvantages: Low efficiency (15%); Very high initial costs; lack of adequate storage materials (batteries); High cost to the consumer.

Hydro:

Advantages: No pollution; Very high efficiency (80%); little waste heat; low cost per KWH; can adjust KWH output to peak loads; recreation money can be generated.

Disadvantages: Fishes are endangered, Sediment buildup and dam failure; changes watershed characteristics; alters hydrological cycle.

Wind:

Advantages: Supplemental power in windy areas; best alternative for individual home owner Disadvantages: Highly variable source; relatively low efficiency (30%).

Geothermal:

Advantages: Very high efficiency; low initial costs.

Disadvantages: Non-renewable (more is taken out than can be put in by nature); highly local resource.

Ocean Thermal Energy Conversion:

Advantages: Enormous energy; steady flow for decades; can be used on large scale; exploits natural temperature gradients in the ocean.

Disadvantages: Enormous engineering effort; extremely high cost; Damage to coastal environments possible.

Tidal Energy:

Advantages: Steady source; energy extracted from the potential and kinetic energy of the earth sun-moon system; can exploit bore tides for maximum efficiency.

Disadvantages: Low duty cycle due to intermittent tidal flow; huge modification of coastal environment; very high costs for low duty cycle source.

Nuclear Energy: A debated energy source, of future.

Alternative Sources of Energy with Respect to their Environmental Impact: To combat global warming and the other problems associated with fossil fuels, we must switch to renewable energy sources like sunlight, wind, and biomass. All renewable energy technologies are not appropriate to all applications or locations and must be chosen after due consideration of all the factors involved. However, as with conventional energy production, there are environmental issues to be considered.

Energy from Wind: It can be taken as a source of energy most benign to the environment. It produces no air or water pollution, involves no toxic or hazardous substances (other than those commonly involved with large machines), and poses no threat to public safety. One of the most misunderstood aspects of wind power is its use of land. Most studies assume that wind turbines will be spaced a certain distance apart and that all of the land in between should be regarded as occupied. This leads to some quite disturbing estimates of the land area required to produce substantial quantities of wind power. According to one widely circulated report from the 1970s, generating 20 per cent of US electricity from windy areas in 1975 would have required siting turbines on 18,000 square miles, or an area about 7 per cent the size of Texas.

In reality, however, the wind turbines themselves occupy only a small fraction of this land area, and the rest can be used for other purposes or left in its natural state. For this reason, wind power development is ideally suited to farming areas. In Europe, farmers plant right up to the base of turbine towers, while in California cows can be seen peacefully grazing in their shadow. The leasing of land for wind turbines, far from interfering with farm operations, can bring substantial benefits to landowners in the form of increased income and land values. Perhaps the greatest potential for wind power development is in the coastal areas like Tamil Nadu, Gujarat, etc. In other settings, however, wind power development can create serious land-use conflicts.

In forested areas it may mean clearing trees and cutting roads, a prospect that is sure to generate controversy, except possibly in areas where heavy logging has already occurred. And near populated areas, wind projects often run into stiff opposition from people. In California, bird deaths from electrocution or collisions with spinning rotors have emerged as a problem in a wind "farm,". In appropriate areas, and with imagination, careful planning, and early contacts between the wind industry, environmental groups, and affected communities, siting and environmental problems can be easily solved.

Solar Energy: Since solar power systems generate no air pollution during operation, the primary environmental, health, and safety issues involve how they are manufactured, installed, and ultimately disposed of Energy is required to manufacture and install solar components, and any fossil fuels used for this purpose will generate emissions.

Thus, an important question is how much fossil energy input is required for solar systems compared to the fossil energy consumed by comparable conventional energy systems. Although this varies depending upon the technology and climate, the energy balance is generally favourable to solar systems in applications where they are cost-effective, and it is improving with each successive generation of technology. According to some studies, for example, solar water heaters increase the amount of hot water generated per unit of fossil energy invested by at least a factor of two compared to natural gas water heating and by at least a factor of eight compared to electric water heating.

Materials used in some solar systems can create health and safety hazards for workers and anyone else coming into contact with them. In particular, the manufacturing of photovoltaic cells often requires hazardous materials such as arsenic and cadmium. Even relatively inert silicon, a major material used in solar cells can be hazardous to workers if it is breathed in as dust. Workers involved in manufacturing photovoltaic modules and components must consequently be protected from exposure to these materials.

None of these potential hazards is much different in quality or magnitude from the innumerable hazards people face routinely in an industrial society. Through effective regulation, the dangers can very likely be kept at a very low level.

The large amount of land required for utility-scale solar power plants-approximately one square kilometer for every 20-60 megawatts (MW) generated-poses an additional problem, especially where wildlife protection is a concern. But this problem is not unique to solar power plants. Generating electricity from coal actually requires as much or more land per unit of energy delivered if the land used in strip mining is taken into account. Solar-thermal plants (like most conventional power plants) also require cooling water, which may be costly or scarce in desert areas.

Large central power plants are not the only option for generating energy from sunlight, however, and are probably among the least promising. Because sunlight is dispersed, small-scale, dispersed applications are a better match to the resource. They can't take advantage of unused space on the roofs of homes and buildings and in urban and industrial lots.

Geothermal Energy: Geothermal energy is heat contained below the earth's surface. The only type of geothermal energy that has been widely developed is hydrothermal energy, which consists of trapped hot water or steam. However, new technologies are being developed to exploit hot dry rock (accessed by drilling deep into rock), geopressured resources (pressurized brine mixed with methane), and magma.

The various geothermal resource types differ in many respects, but they raise a common set of environmental issues. Air and water pollution are two leading concerns, along with the safe disposal of hazardous waste, siting, and land subsidence. Since these resources would be exploited in a highly centralized fashion, reducing their environmental impacts to an acceptable level should be relatively easy. But it will always be difficult to site plants in scenic or otherwise environmentally sensitive areas due to strong public opposition.

The method used to convert geothermal steam or hot water to electricity directly affects the amount of waste generated. Closed-loop systems are almost totally benign, since gases or fluids removed from the well are not exposed to the atmosphere and are usually injected back into the ground after giving up their heat. Although this technology is more expensive than conventional open-loop systems, in some cases it may reduce scrubber and solid waste disposal costs enough to provide a significant economic advantage.

Open-loop systems, on the other hand, can generate large amounts of solid wastes as well as noxious fumes. Metals, minerals, and gases leach out into the geothermal steam or hot water as it passes through the rocks. The large amounts of chemicals released when geothermal fields are tapped for commercial production can be hazardous or objectionable to people living and working nearby.

In large geothermal development sites, steam vented at the surface contains hydrogen sulphide (H_2S)-accounting for the area's "rotten egg" smell-as well as ammonia, methane, and carbon dioxide. At hydrothermal plants carbon dioxide is expected to make up about 10 per cent of the gases trapped in geopressured brines. For each kilowatt-hour of electricity generated, however, the amount of carbon dioxide emitted is still only about 5 per cent of the amount emitted by a coal- or oil-fired power plant. Scrubbers reduce air emissions but produce a watery sludge high in sulphur and vanadium, a heavy metal that can be toxic in high concentrations. Additional sludge is generated when hydrothermal steam is condensed, causing the dissolved solids to precipitate out. This sludge is generally high in silica compounds, chlorides, arsenic, mercury, nickel, and other toxic heavy metals. One costly method of waste disposal involves drying it as thoroughly as possible and shipping it to licensed hazardous waste sites. Research under way at New York points to the possibility of treating these wastes with microbes designed to recover commercially valuable metals while rendering the waste nontoxic.

Usually the best disposal method is to inject liquid wastes or redissolve the solids to pump them back into a porous stratum of a geothermal well. This technique is especially important at geopressured power plants because of the sheer volume of wastes they produce each day. Wastes must be injected well below fresh water aquifers to make certain that there is no communication between the usable water and waste-water strata.

In addition to providing safe waste disposal, injection may also help prevent land subsidence. At Wairakei, New Zealand, where wastes and condensates were not injected for many years. This resulted in sinking of an area by 7.5 meters since 1958.

Most geothermal power plants will require a large amount of water for cooling or other purposes. In places where water is in short supply, this need could raise conflicts with other users for water resources.

The development of hydrothermal energy faces a special problem. Many hydrothermal reservoirs are located in or near wilderness areas of great natural beauty such as Yellowstone National Park in United States and in Laddakh in India. Proposed developments in such areas in USA have aroused intense opposition.

Biomass: Biomass power, derived from the burning of plant matter, raises more serious environmental issues than any other renewable resource except hydropower. Combustion of biomass and biomass-derived fuels produces air pollution; beyond this, there are concerns about the impacts of using land to grow energy crops. How serious these impacts are will depend on how carefully the resource is managed. The picture is further complicated because there is no single biomass technology, but rather a wide variety of production and conversion methods, each with different environmental impacts.

Cause of Air Pollution because of use of Biomass as Fuel: Inevitably, the combustion of biomass produces air pollutants, including carbon monoxide, nitrogen Oxides, and particulates such as soot and ash. The amount of pollution emitted per unit of energy generated varies widely by technology, with wood-burning stoves and fireplaces generally the worst of fenders. Modern, enclosed fireplaces and wood stoves pollute much less than traditional, open fireplaces for the simple reason that they are more efficient. Specialized pollution control devices such as electrostatic precipitators (to remove particulates) are available, but their use will increase the production costs.

Emissions from conventional biomass-fueled power plants are generally similar to emissions from coal-fired power plants, with the notable difference that biomass facilities produce very little sulphur dioxide or toxic metals (cadmium, mercury, and others). The most serious problem is their particulate emissions, which must be controlled with special devices. More advanced technologies, such as the gasifier/combustion turbine combination, should generate much lower emissions, perhaps comparable to those of power plants fueled by natural gas.

Facilities that burn raw municipal waste present a unique pollution-control problem. This waste often contains toxic metals, chlorinated compounds, and plastics, which generate harmful emissions. Since this problem is much less severe in facilities burning refuse-derived fuel (RDF) -pelletized or shredded paper and other waste with most inorganic material removed most waste-to-energy plants built in the future are likely to use this fuel. Co-firing RDF in coal-fired power plants may provide an inexpensive way to reduce coal emissions without having to build new power plants.

Using biomass-derived methanol and ethanol as vehicle fuels, instead of conventional gasoline, could substantially reduce some types of pollution from automobiles. Both methanol and ethanol evaporate more slowly than gasoline, thus helping to reduce evaporative emissions of volatile organic compounds (VOCs), which react with heat and sunlight to generate ground-level ozone (a component of smog). According to US EPA estimates, in cars specifically designed to burn pure methanol or ethanol, VOC emissions from the tail pipe could be reduced 85 to 95 per cent, while carbon monoxide emissions could be reduced 30 to 90 per cent. However, emissions of nitrogen oxides, a source of acid precipitation, would not change significantly compared to gasoline-powered vehicles.

Some studies have indicated that the use of fuel alcohol increases emissions of formaldehyde and other aldehydes, compounds identified as potential carcinogens. Others counter that these results consider only tailpipe emissions, whereas VOCs, another significant pathway of aldehyde formation, are much lower in alcohol-burning vehicles. On balance, methanol vehicles would therefore decrease ozone levels. Overall, however, alcohol-fueled cars will not solve air pollution problems in dense urban areas, where electric cars or fuel cells represent better solutions.

Greenhouse Gases and Biomass Energy: A major benefit of substituting biomass for fossil fuels is that, if done in a sustainable fashion, it would greatly reduce emissions of greenhouses gases. The amount of carbon dioxide released when biomass is burned is very nearly the same as the amount required to replenish the plants grown to produce the biomass. Thus, in a sustainable fuel cycle, there would be no net emissions of carbon dioxide, although some fossil-fuel inputs may be required for planting, harvesting, transporting, and processing biomass. Yet, if efficient cultivation and conversion processes are used, the resulting emissions should be small (around 20 per cent of the emissions created by fossil fuels alone). And if the energy needed to produce and process biomass came from renewable sources in the first place, the net contribution to global warming would be zero.

Similarly, if biomass wastes such as crop residues or municipal solid wastes are used for energy, there should be lesser or no net greenhouse gas emissions. This will also reduce the release of methane from landfills.

Implications for Agriculture and Forestry: One effect of growing trees and other plants for energy is that it could benefit soil quality and farm economies. Energy crops could provide a steady supplemental income for farmers in off-seasons or allow them to work unused land without requiring additional equipment. Moreover, energy crops could be used to stabilize cropland or rangeland prone to erosion and flooding. Trees would be grown for several years before being harvested, and their roots and leaf litter could help stabilize the soil. The planting of coppicing, or self-regenerating, varieties like eucalyptus, poplar, etc. would minimize the need for disruptive tilling and planting. Perennial grasses harvested like hay could play a similar role.

If improperly managed, however, energy farming could have harmful environmental impacts. Although energy crops could be grown with less pesticide and fertilizer than conventional food crops, large-scale energy farming could nevertheless lead to increases in chemical use simply because more land would be under cultivation. It could also affect biodiversity through the destruction of species habitats, especially if forests are more intensively managed. If agricultural or forestry wastes and residues were used for fuel, then soils could be depleted of organic content and nutrients unless care was taken to leave enough wastes behind. These concerns point out the need for regulation and monitoring of energy crop development and waste use.

Energy farms may present a perfect opportunity to promote low-impact sustainable agriculture, or, as it is sometimes called, organic farming. Such methods could be adapted to energy farming. Nitrogen-fixing crops could be used to provide natural fertilizer, while crop diversity and use of pest parasites and predators could reduce pesticide use. Though such practices may not produce as high a yield as more intensive methods, this penalty could be offset by reduced energy and chemical costs. Increasing the amount of forest wood harvested for energy could have both positive and negative effects. On one hand, it could provide an incentive for the forest-products industry to manage its resources more efficiently, and thus improve forest health. But it could also provide an excuse, under the "green" mantle, to exploit forests in an unsustainable fashion. Unfortunately, commercial forests have not always been soundly managed, and many people view with alarm the prospect of increased wood cutting. Their concerns can be met by tighter government controls on forestry practices and by following the principles of "excellent" forestry. If such principles are applied, it should be possible to extract energy from forests indefinitely.

Hydropower: The development of hydropower has become increasingly problematic due to strict considerations being given to adverse environmental impacts. The biggest obstacle in construction of large dams is that most suitable undeveloped sites are very often reserve forest, or inhabited by people. The Silent Valley project in Kerala was forced to be abandoned by public opposition. In United States for example, a series of large facilities on the Columbia River in Washington was forced to reduce its peak output by 1,000 MW to save an endangered species of salmon. Salmon numbers have declined rapidly because the young are forced to make a long and arduous trip downstream through several power plants, carrying risk of death from turbine blades at each stage. To ease this trip, hydropower plants may be required to divert water around their turbines at those times of the year when the fish attempt the trip.

That environmental concerns would constrain hydropower development more and more difficult in future though these plants produce no air pollution or greenhouse gases. Yet, as the salmon example makes clear, they affect the environment.

The impact of very large dams is so great that many developed countries had dropped the plans to construct very large dams though they are being constructed in many developing countries like India and China due to their lucrative economics. The reservoirs created by such projects frequently inundate large areas of forest, farmland, wildlife habitats, scenic areas, and even towns.

In addition, the dams can cause radical changes in river ecosystems both upstream and downstream. Small hydropower plants using reservoirs can cause similar types of damage, though obviously on a smaller scale. Some of the impacts on fish can be mitigated by installing "ladders" or other devices to allow fish to migrate over dams, and by maintaining minimum river-flow rates; screens can also be installed to keep fish away from turbine blades.

In one case, flashing underwater lights placed in the Susquehanna River in Pennsylvania direct night-migrating American shad around turbines at a hydroelectric station. As environmental regulations have become more stringent, developing cost-effective mitigation measures such as these is essential. It is also a fact that hydropower is almost certainly approaching the limit of its potential in most of the developed countries.

Formation of Earth

Most soils are composed of small particles of weathered rock that are silicate minerals. These minerals consist of polymeric structures in which the fundamental unit is a silicon atom surrounded tetrahedrally by four oxygen atoms, the latter in turn are each bonded to another silicon, and so on, the resulting structure is an extended network. There are variations in the structure and H^+ Na^+, K^+, Mg^{2+}, Ca^{2+}, Al^{3+} Fe^{2+} may be in the structure. The weathering of the silicate minerals from rocks can involve chemical reactions such as the substitution by aluminium for some of the silicon atoms. Eventually such reactions yield clay minerals which are a class of soil materials. Clay minerals have a particle size less than 2 mm. Small colloidal clay particles possess an outer layer of cations which are bound electrostatically to an electrically-charged inner layer.

Depending on the concentration of cations in the water surrounding the clay particle, the cations on the particle are capable of being exchanged, *e.g.* if the soil is acidic, the metal ions on the surface will be displaced by H^+ ions and the metal ions will then enter the aqueous phase.

Besides minerals soil contains organic matter, water and air in varying concentrations. The organic matter, which gives soil its dark colour is primarily a material called humus, which is undecomposed plant material. It contains mostly protein and lignin, both of which are mostly insoluble in water. Lignin contains polymeric strands of benzene rings connected by chains of carbon and oxygen atoms. Due to partial oxidation of some of the lignin, many of the resulting polymeric strands contain Carboxylic acid groups, - COOH; this portion of humus is known as humic and fulvic acids, and is soluble in alkaline solutions.

The ultimate sink for heavy metals, and for many toxic organic compounds as well, is deposition and burial in soil and sediments. Heavy metals often accumulate in the top layer of the soil, and are therefore accessible for uptake by the roots of crops.

Humic materials have a great affinity for heavy metal cations, and extract them from the water that passes through them. This happens mostly because of the formation of complexes with the metal ions by - COOH groups in the humic and fulvic acids.

Humic acid complexes are normally water - insoluble and smaller fulvic acid complexes are water-soluble.

Heavy metals are retained by soil in three ways:

(a) By adsorption onto the surfaces of mineral particles

(b) By complexation by humic substances in organic particles

(c) By precipitation reactions.

Like many other chemicals, heavy metals often are adsorbed onto the surfaces of particulates - especially organic ones—suspended in water. In lakes such particles eventually settle to the bottom of the lakes and are buried when other sediments accumulate on top of them. Sometimes such sink mechanism doesn't work due to desorption of chemicals or consumption of particles by bottom-growing organisms or due to anaerobic bacterial activities. Ex. Anaerobic bacteria methylate the mercuric ion to form $(CH_3)2Hg$ and CH_3HgX, which then rapidly desorb from sediment particles and dissolve in water, and enter in food chain. The biogeochemical cycle of bacterial methylation and demethylation of mercury in sediments.

Analysing the Risk: The objective of risk analysis is to provide a scientific framework to help decision makers and other concerned individuals in making judicious decisions that will ultimately solve or mitigate health and environmental problems. Risk analyses are used in many situations like

(a) Analyzing potential health and environmental problems resulting from exposure to chemicals

(b) Selecting sites for hazardous-chemical-producing units

(c) Comparing alternate technologies available

(d) Analyzing potential risks due to fire and explosion hazards.

To be meaningful and cheap risk analyses are usually based on a relatively narrow set of hazards *i.e.,* they focus on a particular problem, for example, cancer and chronic health risks while not considering others like ecological problems.

The terms and their definitions used in the field of risk analysis are given below:

Hazard: Hazard refers to the source of risk. For example a toxic chemical is a hazardous to human health but is not a risk unless someone is exposed to it.

Hazard Identification: It is the first step in risk analysis. It is the process of reconising all the hazards with the potential to harm an individual or the environment.

Risk: It is the probability of exposure of individuals, populations or ecosystems to toxic substances or to hazardous conditions. Risks can be lowered by reducing any of its components.

(a) The probability of risk taking place

(b) The probability of exposure

(c) The severity of consequences.

Risk Assessment: It is the second step in risk analysis. It is an estimate of the severity and likelihood of harm to populations or ecosystems from exposure to a hazard. It includes

(a) Assessment of sources

(b) Assessment of exposure

(c) Assessment of dose-response

(d) Risk characterization, which calculates a numerical estimate of the risk.

Risk management: It is the effective use of a risk assessment. It may include evaluation of alternatives, informed decisions and the follow through activities.

Analysing the Risk: Risk analysis includes four steps:

(a) hazard identification

(b) risk assessment

(c) risk management

(d) risk communication

Risk analyses use science, engineering, probability and statistics to estimate and evaluate the magnitude and likelihood of health and environmental risks.

(a) *Hazard Identification:* It establishes the existence of hazards at specific site. Hazards may be toxic substances or specific conditions that have the potential to be harmful to human health or the environment.

(b) *Source:* For risk analysis further information like, toxicity, rate of discharge, conditions of release (temperature and pressure), etc. are required to decide a worst case scenario.

(c) *Exposure Assessment:* It determines the identification of the toxic species, frequency of exposure, duration of exposure and the routes of exposure to human populations and the environment. Exposure assessment data is obtained from actual monitoring, mathematical models, or reasonable scientific estimates.

(d) *Dose-response Assessment:* This is also called toxicity assessment. It involves

(i) identification of the adverse health effects due to chemical exposure

(ii) relationship between the magnitude of the exposure and adverse effects.

The response to an exposure may be

(i) no observable effect

(ii) temporary effects, like eye irritation

(iii) permanent injury like liver damage

(iv) chronic functional impairment like emphysema of lungs

(v) death

(e) *Risk Characterization:* In it the results of exposure assessment and dose-response are used to determine if the current or future levels of exposure are potentially problematic. Toxicology and exposure results are combined to give qualitative and quantitative expressions of risk. Risk characterization includes the probability of health and environmental effects.

(f) *Risk Management:* It uses all the informations discussed above, together with other technical, business, environmental, social and political information to

(i) evaluate alternative

(ii) determine the best action to reduce or eliminate risk.

(iii) design a plan to implement this action.

(g) *Risk Communication:* The results of a risk analysis must be communicated to technical and public communities to gain benefits. The purpose of an effective risk analysis is to assist decision makers, companies and citizens to work on things that matter most and ignore things that matter the least.

Desertification: It is a process where by the productive potential of arid and semi and land falls by 10% or more. Moderate desertification is 10-25% drop in productivity, severe desertification is a 2550% drop, and very severe desertification is a drop of 50% or more, generally creating huge gullies and sand dunes.

Since last 50 years, an estimated 8.1 million square kilometers have become desertified. This threatens the livelihoods of at least 900 million people in 100 countries. Every year low-to-moderate desertification affects 60,000 square kilometer area, and another 210,000 square kilometer area undergo severe desertification and loose so much soil and fertility that they are no longer useful for farming or grazing.

Reasons for Desertification

1. Overgrazing on fragile arid and semi and rangelands.
2. Deforestation without reforestation.
3. Surface mining without land reclamation.
4. Build-up of salt and water logging of soil.
5. Farming on land with unsuitable terrain or soil.
6. Soil compaction by heavy farm machinery and cattle hoof.

- *Desertification in hilly regions of the Himalayas:* Deforestation in Himalayas, involving clearance of natural forests and plantation of monocultures like *Pinus roxburghi, Eucalyptus camaldulensis,* etc. have upset the ecosystem by changing various soil (edaphic) and biological properties. Nutrient cycling has become poor, original rich germplasm is lost and the area is invaded by exotic weeds. These areas are not able to recover and are losing their fertility. The entire west Khasi hill district of Meghalaya in North-east Himalayas, Laddakh and parts of Kumaon and Garhwal are now facing the serious problem of desertification.
- *Disappearing Tea gardens in Chhota Nagpur:* This hilly region used to be a good forested area towards the turn of the century and used to receive fairly frequent afternoon showers favouring tea plantations. Following the destruction of forests, rainfall declined in Chhota Nagpur to such an extent that tea-gardens also disappeared from the region.
- *Waning Rainfall in Udhagamandalam (Ooty):* The sub-normal rainfall during 1965-84 at Ooty in Nilgiri mountains has been found to be closely associated with declining forest cover in this region

in the past 20 years. The rainfall pattern was found to fluctuate with wooded land area in the hills. When the Nilgiri mountains had luxuriant forest cover annual rainfall used to be much higher.

Sardar Sarovar Dam (Uprooted Forests And Tribals)

A Case Study: The dam is situated on river Narmada and is spread over three states of Gujarat, Maharashtra and Madhya Pradesh. Although the project is aimed at providing irrigation water, drinking water and electricity to the three states, the environmental impacts of the project have raised challenging questions. A total of 1,44,731 ha of land will be submerged by the dam, out of which 56,547 ha is forest land. A total of 573 villages are to be submerged by the Narmada Dam.

Submergence of about 40,000 ha of forest under Narmada Sagar, 13,800 ha under Sardar Sarovar and 2,500 ha under Omkareshwar would further create pressure on remaining forest areas in adjoining areas. Submergence area is very rich in wildlife *e.g.* tigers, panthers, bears, wolves, pangolins, hyenas, jackals, flying squirrels, antelopes, black bucks, chinkara, marsh crocodiles, turtles, etc. Many of these species are listed in schedule I & II of Wildlife Protection Act, 1972. Thus massive loss of these wildlife species is apprehended, due to the devastation of the forest under the project.

As per the estimates of the Institute of Urban Affairs, New Delhi, the Narmada valley project will lead to eventual displacement of more than one million people, which is probably the largest rehabilitation issue ever encountered as per the World Bank. Uprooting of the tribals and their forced shifting in far-flung areas may not be easily adjusted to.

Besides serious economic deprivation, the displacement will affect the tribal peoples' culture, their beliefs, myths and rituals, festivals, songs and dances, all closely associated with the hills, forest and streams. Most of these tribals belong to poor, under privileged schedule castes and tribes who are being uprooted from a place where they have lived for generations. The displaced persons have to undergo hardship and distress for the sake of development and prosperity of a larger section of the society. It is therefore the duty of the project proponents and government to pay maximum attention for proper rehabilitation of the displaced tribals.

Case Studies

• *Mining and Quarrying in Udaipur*

About 200 open cast mining and quarrying centers in Udaipur, about half of which are illegal are involved in stone-mining including soapstone, building stone, rock phosphate and dolomite. The mines spread over 15,000 hectares in Udaipur have caused many adverse impacts on environment. About 150 tonnes of explosives are used per month in blasting. The overburden, wash-off, discharge of mine water, etc. pollute the water. The Maton mines have badly polluted the Ahar river. The hills around the mines are devoid of any vegetation except a few scattered patches and the hills are suffering from acute soil erosion. The waste water flows towards a big tank of "Bag Dara". Due to scarcity of water people are compelled to use this effluent for irrigation purpose.

The blasting activity has adversely affected the fauna and the animals like tiger, lion, deer and even hare, fox, wild cats and birds have disappeared from the mining area.

• *Mining in Sariska Tiger Reserve in Aravallis*

The Aravalli range is spread over about 692 km in the North-west India covering Gujarat, Rajasthan, Haryana and Delhi. The hill region is very rich in biodiversity as well as mineral resources. The Sariska tiger reserve has gentle sloppy hills, vertical rocky valleys, flat plains as well as deep gorges. The reserve is very rich in wildlife and has enormous mineral reserves like quartzite, Schists, marble and granite in abundance. Mining operations within and around the Sariska Tiger reserve has left many areas permanently infertile and barren. The precious wildlife is under serious threat. We must preserve the Aravalli series as a National Heritage and the Supreme Court on December 31st, 1991 has given a judgement in response to a Public Interest Litigation of Tarun Bharat Sangh, an NGO wherein both Centre and State Government of Rajasthan have been directed to ensure that all mining activity within the park be stopped. More than 400 mines were shut immediately. But, still some illegal mining is in progress.

- ***Uranium Mining in Nalgonda, A.P.—The public hearing***

The present reserves of Uranium in Jaduguda mines, Jharkhand can supply the yellow cake only till 2004. There is a pressing need for mining more uranium to meet the demands of India's nuclear programme. The Uranium Corporation of India (UCIL) proposes to mine uranium from the deposits in Lambapur and Peddagattu villages of Nalgonda district in Andhra Pradesh and a processing unit at about 18 kms at Mallapur. The plan is to extract the ore of 11.02 million tons in 20 years.

The IUCL is trying its best to allure the villagers through employment opportunities. But, experts charge the company for keeping silence on the possible contamination of water bodies in the area. The proposed mines are just 1 km from human habitation and hardly 10 km from Nagarjun Sagar Dam and barely 4 km from the Akkampalli reservoir which is Hyderabad's new source of drinking water.

It is estimated that 20 years of mining would generate about 7.5 million metric tones of radioactive waste of which 99.9% will be left behind. The villagers are very likely to be affected by the radioactive wastes. Though IUCL claims that there won't be any such accidents, but no one can deny that it is a highly hazardous industry and safety measures cannot be overlooked. The pathetic condition of Jaduguda Uranium mines in Jharkhand where there is a black history of massive deaths and devastation have outraged the public, who don't want it to be repeated for Nalgonda. The proposed mines would cover about 445 ha of Yellapurum Reserve Forest and the Rajiv Gandhi Tiger Sanctuary. The public hearing held just recently in February, 2004 witnessed strong protests from NGOs and many villagers. The fate of the proposed mining is yet to be decided.

Case Studies: Salinity and water logging in Punjab, Haryana and Rajasthan:

The first alarming report of salt-affected wasteland formation in connection with irrigation practices came from Haryana (then Punjab) in 1858. It was reported that several villages in Panipat, Rohtak and Delhi lying in command area of Western Yamuna Canal were suffering from destructive saline efflorescence. The "Reh Committee" in 1886 drew the attention of the government on some vital points showing a close relationship between irrigation, drainage and spread of "reh" and "usar" soils.

The floods of 1947, 1950, 1952, 1954-56 in Punjab resulted in aggravated water logging with serious drainage problems. Introduction of canal irrigation in 1.2 m ha in Haryana resulted in rise in water-table followed by water-logging and salinity in many irrigated areas causing huge economic losses as a result of fall in crop productivity. Rajasthan too has suffered badly in this regard following the biggest irrigation project "Indira Gandhi Canal Project" and the sufferings of a big area in Western Rajasthan have changed from a condition of "water-starved wasteland" to that of a "water soaked wasteland".

Case Study: A build-up of DDT concentration: A striking case of biomagnification of DDT (a broad range insecticide) was observed when some birds like Osprey were found to suffer a sharp decline in their population. The young ones of these birds were found to hatch out in premature condition leading to their death. This was later found to be due to biomagnification of DDT through the food chain. DDT sprayed for pest control was in very low concentration, but its concentration increased along the food chain through phytoplanktons to zooplanktons and then to fish which was eaten by the birds.

The concentration of DDT was magnified several thousand times in the birds which caused thinning of shells in the birds' eggs, causing death of the young ones. It becomes very clear from the above instance that the animals occupying the higher trophic levels are at a greater risk of biomagnification of toxic chemicals. Human beings consuming milk, eggs and meat are at a higher trophic level. So, we have to stop indiscriminate use of pesticides and heavy metals if we wish to save ourselves from their biologically-magnified toxic levels.

2

Rapid Development

In our country Science is witnessed to be taught from very early days. If we look at the development of science in India, we find that very early in her civilisation India developed a great interest in mathematics and Ayurveda. During British Empire introduction of modern science in India was extremely slow, and the development of science in India was greatly accelerated after independence.

The Background

Before the rise of modern science in Europe, around the 17th century, the level of advancement in sciences (astronomy, mathematics, medicine, biology, metaphysics, etc.) achieved by ancient and medieval societies in the old world did not differ appreciably from one cultural area to another.

In India the development of sciences is as old as her civilisation itself. Her peculiar geographical position enabled her to become the natural meeting ground of many nations and cultures and, in consequence, played an important role in the transmission and diffusion of ideas very early in her civilisation, India developed medical and alchemical practices of the Ayurveda. In mathematics, India developed a great interest and aptitude and made notable contributions to the number theory, the decimal place value, algorism, trigonometry and algebra.

Gradual Cultivation

In the beginning of the nineteenth century people with varying degrees of scientific background (medical men, naturalists and engineers) started coming to India from Europe, at first in search of services under local princes and chieftains and later on to man the scientific surveys and establishments set up by the government. These men framed in European institutions and laboratories, spent the best part of their lives in India and left an excellent record of their work in various branches of science.

Despite such long contacts, introduction of modern science in India was extremely slow. There were many causes of this delayed reaction. The European scientific world in India was limited to field sciences and not to basic sciences that depended on mathematics and laboratory work. Indian fauna and flora attracted the attention of European naturalists from the seventeenth century. Modern zoological researchers in India had their beginnings in random and scattered observations by the naturalists on elephants, fishes, serpents, molluscs, birds and mammals.

The Colonial Government adopted from the beginning a policy of secrecy and exclusion of Indians from Government scientific work. For technological jobs not considered of military importance, restrictions against the employment of Indians were relaxed as they could be employed very cheaply. As an example of the low salaries to trained and qualified Indians/ the order of the Governor-General dated 28th January 1835, establishing the Calcutta Medical College, regulated that "as inducement for pupils of a respectable class to enter the Institution, the pay of the Native Doctors, who shall have been educated at the College, and have received the certificates of qualifications, shall be Rs. 30.00 p.m. rising to Rs. 50.00 p.m. after 14 years of service, whereas an European Assistant to the Superintendent of the college shall draw a staff salary of Rs. 600.00 p.m. in addition to his registered pay and allowances." Calcutta Medical College established in 1835 under a number of capable teachers, became an important institution, for the study of anatomy, physiology and medicine (along with surgery), as well as of chemistry, botany and natural philosophy.

Indians asked for Western Education in Science from the Government. The Colonial power agreed to create facilities but with miserly financial provision just sufficient to train a few clerks able to operate in English in Government offices and European Commercial houses, and not to enable the native people to pick up treasures from European science. So the hard way of working for science began. Mahendra Lal Sircar *(born* in 1833), an M.D. of the Calcutta Medical College, clearly saw that science would never develop and strike deep roots in this country under foreign tutelage and that Indians themselves must come forward to raise funds and found institutions for the training of scientists and organisation of research.

In 1876 Dr. Sircar himself founded India's first research institution, "The Indian Association for the Cultivation of Science," completely under Indian Management and control with finances derived from private subscriptions. In the first half of the twentieth century the association developed into an important centre for research in the physical sciences—in optics, acoustics, scattering of light, X-rays and magnetism; and C.V. Raman performed his Nobel Prize Winning Experiments in its unassuming laboratories, *i.e.* western India. Jamshedji Tata a businessman, prepared plans for a similar institution for technical and scientific education and research which finally took the shape of "The Institute of Science" at Bangalore in the beginning of the twentieth century.

The Educational Despatch of 1854 paved the way for University education. In 1857 the three Universities of Calcutta, Bombay and Madras came into existence, but they came out just as examining bodies with powers to grant degrees. The actual teaching and academic work were left to the colleges. Fortified by the University Act of 1904, which empowered Universities to appoint professors and lecturers, to hold and manage educational endowments, and to erect, equip and maintain libraries, laboratories and museums, Asutosh Mookerji, Vice Chancellor of Calcutta University took the initiative in establishing the, first University College of Science. Undaunted by the Government's refusal to provide funds for the creation of professorships and other facilities, Asutosh obtained princely endowments from Sir Tarak Nath Palit and Sir Rash Behary Ghosh, who had amassed enormous amounts of money and property in the legal profession, and established a number of chairs in Chemistry, Physics, Applied Mathematics and Botany.

According to the terms of the endowment, professorships could be filled only by Indians, a clause which further irritated the Government firmly entrenched in the view that only Europeans were suited for such high positions. This University College of Science, although starved financially all through, produced a group of physicists and chemists who received international recognition for their scientific developments and institutions staffed by high salaried Europeans and maintained and patronised by Government funds cut a sorry figure.

The development of science in India was greatly accelerated after independence (August 1947). In 1950 the Government of India appointed a Planning Commission for preparing a blueprint of all-round economic development. In 1954 the Indian Parliament accepted socialism as a political goal. Declaring these objectives, fullest emphasis was laid on the development of science and technology on all fronts. In 1957, the Government took one step further in adopting a National Science Policy Resolution that envisaged the cultivation of science and scientific research in all its aspects, assured an adequate supply, within the country, of research scientists of highest quality through an intensive programme of training, promised the availability of conditions and an atmosphere of academic freedom in which the creative talent of men and women would find full scope in scientific activity. The resolution thus reaffirmed the Government decision to encourage science and develop a healthy scientific community as a sound basis after a balanced economic development.

Role of Science

- In the beginning of twentieth century science was not a school subject in our country. The Report of the Secondary School Commission 1953, recommended the teaching of General Science as a compulsory subject in the high and higher secondary schools.
- The All India Seminar on the Teaching of Science in Secondary Schools (1956) dealt with almost all the problems facing the inclusion of General Science as a Core Subject for the Higher Secondary Classes—syllabus, apparatus, teaching aids, textbooks, science clubs, science museums, examination techniques etc. It suggested a uniform system of science teaching for the entire country, suited to its needs and resources.
- Indian Parliamentary and Scientific Committee (1962) studied the allied problems of science education in schools like:–growth of school population,–shortage of qualified science teachers,–accelerated achievement in science,–demand for increase in technically trained manpower,–growing importance of science in the affairs of mankind, and–changes in the processes and goals of science.

- UNESCO Planning Commission (1963-64) worked on the problems of science education in India and suggested ways to improve it. As a follow up programme, Department of Science and Mathematics Education of the NCERT took up the pilot project of preparing new disciplinary science curricula at middle (VI-VIII) level -text books (Physics, Chemistry, Biology), teachers' guides, science kits, kit guides, teacher training films and evaluation material. This Disciplined Science Programme was changed to Integrated Science Programme in mid-eighties.
- In 1970 under UNICEF Assisted Science Education Programme (SEP) Primary "Science is Doing" Programme was developed by NCERT, which was used in primary schools throughout the country. This Programme was a package of classes I and II Science Syllabus, Class III-V Science Texts, Teachers Guides, Primary Science Kit, Kit Guide, and Teacher Training Films. This Programme was changed to EVS Programme in mid-eighties.
- Kothari Commission (1964-66) recommendations were implemented in 1975 when Science for All (SFA) was introduced as a part of general education during the first ten years of schooling. With this 10+2+3 education scheme started with an additional year of schooling, in the country. First Disciplined Science Course (Physics, Chemistry, Biology) was introduced at secondary level (IX-X). This was changed to Disciplined Science A-Course (Physics, Chemistry, Biology) and Integrated Science B-Course. Students had an option either to take Science A-Course or to take Science B-Course. This again changed to just one Integrated Science Course for all. At +2, Senior Secondary level (XI-XII) Disciplined Science Course (3 different science subjects-

Physics, Chemistry, Biology) started from the very beginning of 10+2+3 education scheme.

- Then in 1986 the National Education Policy Document (NPE-1986) came out. If some one asks, "What is new in this new education policy (NPE-1986), "Implementation" perhaps will be the right answer. Much emphasis has been given to quality pre-service and in-service teacher education in the policy document. For this District Institutes of Education (DIETs) and Institutes of Advance Studies in Education (LASEs) were established throughout the country for Elementary and Secondary Education respectively.

New Approaches

India is now engaged in a broad spectrum of scientific research, both fundamental and applied, in Government, Universities and Private Research Establishments. In recent years there has been extraordinary success in developing new polymers, ceramics, composites, superconductors, nanomaterials, smart materials and biomaterials. Biotechnology, Genetic Engineering and Biomedical Research are some other fields in which India has started entering.

AIDS (Acquired Immune Deficiency Syndrome) poses a threat to India as a large number of people are infected with HIV (Human Immunodeficiency Virus). There are no drugs today for AIDS. The problem which has dogged anti HIV drugs is that resistant mutant forms of virus are formed within mere weeks. Lot of research work is to be done in search of a vaccine against HIV. Per capita consumption of energy in India is very low compared to developed countries, and even that we are unable to afford. In the years to come when we have already entered twenty-first century, we will need much more energy. The major effort in next decade would have to be through an increase in the production of coal and a search of new reserves of oil. We should also give due emphasis to new technologies for solar energy and hydrogen energy. India is also to expand its nuclear power programmes. Nuclear energy could play an important role in meeting our bulk energy requirement. Our expertise in the field of nuclear power technology as well as related research areas, is an asset for us.

For all this we need huge amounts of funds, as scientific research has become costly. If we compare the funds available for scientific research in India compared to some other developing countries, India has a very gloomy picture. South Korea had planned to increase its R&D spending to over 5 per cent of its GDP by year 2000.

China had planned to increase its R&D investment from 0.5 per cent to 1.5 per cent of GDP by the year 2000. Unfortunately India's R&D expenditure has come down to 0.89 per cent of GDP from 1.1 per cent earlier. We need to realise that in order to be able to do competitive scientific research and development we have to bring our R&D expenditure to at least about 1.5 per cent of GDP and focus our efforts on a few selected programmes and projects.

Our scientists want people to support them and understand their needs, and they are to inform people about science, why science is needed, what they are doing and why. They (our scientists) are to build up partnership with public, industries, politicians and bureaucrats. They need informed friends of science at all places.

When changes which affect our future, are happening and will happen at such a rapid rate, and are based on science and technology, it is necessary that our scientists be more close to decision making.

In the existing state of scientific advancement and development of resources for research, the rate of scientific growth of developed countries is likely to continue to be faster than that of the developing countries like India. Science has become deeply involved with defence and big industry, with the result that big sciences like atomic energy, space, etc., has been and probably will remain concentrated in super powers, In nuclear, space, computer and few other sciences, developing nations like India are already at the mercy of super powers. To be self-sufficient we are to change our science curriculum right from the school stage. We are not just to teach science but also scientific method. We are not just to teach science content but also science processes, the ways in which scientists advance their knowledge and solve problems.

Science should be presented to students as a way in which they can conduct an inquiry into the nature of things as well as a body of information built up by other people. The science processes are being neglected and the school science has been concerned almost exclusively with the content—the body of information. Science is not just content. Science is content plus science processes.

If we want to advance in science like other developed countries, science processes should be given due importance like science content in school science curriculum, and students should be encouraged to become personally involved in solving problems and in discovering some science for themselves.

3

Essential Factors

The terrestrial environment is the most complex type of all, for in terrestrial environment there is a great fluctuation and interaction of environmental factors than in an aquatic medium. The interaction of physical, climate and biotic factors produces a wide variety of ecologic conditions on the continents and islands.

Importance of Air

1. The air is the source of oxygen and carbon dioxide for the living biota in terrestrial habitat. But none of them acts as a limiting factor, because of their great abundance.
2. Air acts as a medium of support for terrestrial biota and is less buoyant than water. An organism on land needs a comparatively rigid skeleton to live and move, whereas in water an organism needs relatively little support.

Importance of Climate

The most important aspects of climate are moisture, temperature and light.

1. *Moisture:* In sharp contrast to aquatic forms, the terrestrial organisms lack constant contact with water. As a result moisture is a major limiting factor on land. The moisture content of air and soil varies from complete saturation (equatorial region) to slight amount (desert) at different places and in different weathers. Terrestrial organisms face the problems of obtaining water and preventing its excessive loss by evaporation and perspiration. Evaporation of water from the bodies of organisms and from the earth surface increases with rise in the relative humidity.
2. *Temperature:* Terrestrial ecosystem is heated by the absorption of heat by earth surface and air because of its low specific heat in comparison to water. Thus heat is gained and lost far rapidly by land than by water. Therefore, fluctuations of temperature diurnally and seasonally are more pronounced on land than in water. It varies from far below freezing point (-68°C) to 60°C in some deserts.
3. *Light of Solar Rradiations:* Solar radiations include radiations of different wavelengths like solar spectrum and ultraviolet and infrared radiations.

Different wavelengths have different effects on organisms. Ultraviolet radiations in large concentrations are harmful to most animals. Green light of the solar spectrum is reflected by plants and the rate of photosynthesis is maximum in red light. Some animals (fish, snails etc.) exhibit an increased growth rate in different wavelengths in the following order-green, red, white, yellow, blue and violet.

In general terrestrial organisms are exposed to much higher intensities of solar radiation than the aquatic organisms and have evolved protective exoskeleton in the form of scales, hair, feathers, cuticle or chitin.

Importance of Soil

In terrestrial ecosystems the soil serves two major functions:

1. It provides support for living organisms and it is a source of all essential nutrients (nitrates, phosphates etc.) for plants. The chemical nature and physical texture of the soil, sand and rocks is particularly vital in the distribution of organisms in terrestrial environment. Different types of soil have different properties and affect the availability of nutrients to plants and thus control the productivity of ecosystem and the nature of biota.
2. The land unlike ocean, is not continuous. Several geographical barriers hinder the free movements of terrestrial organisms.

Role of Gravity

1. Aquatic organisms counteract gravity by evolving various floatation and swimming mechanisms. But a land animal has to support its entire weight. The animals best adapted to terrestrial life have developed appendages in the form of legs or wings for rapid locomotion over the surface or through the air.
2. Fluid buoyance permits water-dwellers to attain huge size, whereas gravity limits the mass of land animals.

The terrestrial biotic communities living in different regions of the earth vary according to temperature, moisture, available mineral matters, light, other forms of life and other limiting ecological factors. This means that regional climates interact with their regional biota and substrata and this results in the organization of distinct easily recognisable community units, called biomes. A biome is a large land community unit, which can be easily recognised by its characteristic flora and fauna.

In each biome the kind of climax vegetation is uniform. Thus the climax vegetation of the grassland biome is grass, conifers in Taiga, deciduous trees in deciduous forest biome and so on, but the particular species of climax plants may vary even in different parts of the same biome.

The kind of climax vegetation depends upon the physical environment. According to plant ecologists a biome is identical with the major plant formation but it includes the animals varieties also and the occurrence of animal varieties is determined by the interaction of climax vegetation and physical forces. The principal biomes of the world are:

1. Tundra,
2. Forest,
3. Grassland,
4. Desert,
5. Mountain and
6. Caves.

Role of Tundra Biome: Tundra Biome lies in fat north between the Arctic ocean and polar ice caps in the north and the coniferous forests to the south. There are two tundra biomes, one in the Palearctic (Europe and Siberia) and the other one in Nearctic region (North America).

Significance of Climate

Tundra biome is characterised by severe winter with a short growing period of 60 days. Mean monthly temperature varies between extremes of –35° (minimum) and +13°C (maximum), the warmest months averaging below 10°C. The annual precipitation is less than 4 cm. Most of the rain comes during summer and early autumn. But humidity is high and evaporation is low.

The ground remains permanently frozen and only the top soil upto a depth of only a few centimeters thaws in summer. The permanently frozen deeper layer is known as permafrost (RAY 1951).

Freezing and thawing lead to the fragmentation of large boulders into small rocks, cause downward slumping of soil slopes and lead to several irregularities in the topography. The general moulding of landscape by frost action is called cryoplanation. This makes the soil unstable and limits the growth and kind of vegetation.

Significance of Vegetation

Tundra is characterised by complete absence of trees. Its vegetation consists of sedges, rushes, lichens, mosses, decumbent shrubs, flowering herbs and grasses. These exhibit some common features. The plants are generally small with stunted growth and compact structure, adapted to desiccation and mechanical abrasion from wind and snow. Most of the plants are perennial because germination of seeds is poor.

Fauna: Inspite of extremely severe environment, the tundra gleems with life. The common mammals are reindeer, caribou, musk ox, arctic fox, arctic hare, arctic shrew, arctic ground squirrel, Tundra vole, lemmings, gray wolf, grizzly bear and polar bear etc. A large number of migratory birds visit Tundra during summer. These are arctic loon, red throated loon, goose, crane, gray falcon, hawks, plovers, gulls, larks, raven, snow bunting, arctic tern, snowy owl and purple sand piper etc. Reptiles and birds are either absent or represented very rarely.

Invertebrate fauna is comparatively restricted in variety. Snails, spiders, mites are well represented. There are swarms of flies and mosquitoes, just a few species of Lepidoptera, Hymenoptera and Coleoptera, whereas species of Hemiptera, Homoptera, Orthoptera and Odonata are almost absent.

Adjustment of Animals

The animals living in Tundra exhibit following adjustments to the severe climatic conditions:

(i) *Loss of Pigmentation:* The animals are mostly white or light gray in colour. This helps them to conceal and to protect against enemy.

(ii) *Hibernation:* Most cold-blooded animals remain inactive during colder months. Invertebrates pass winter in larval or pupil stage resistant to freezing. Rotifers, dysticid and hydrophilid beetles remain frozen in ice for months and resume activity in winter.

(iii) *Insulation:* Warmblooded mammals and birds have thick fur or dense plumage and a thick layer of fat. These measures help in retaining body heat.

(iv) *Hiding or Roosting:* Voles and lemmings etc. escape winter by staying in their nests or tunnels under snow.

(v) *Migration:* Some mammals and majority of birds in winter migrate to southern portions of Tundra or other warmer regions.

Biome of Forests

Forest is a large uncultivated and uninhabited tract of land covered with trees of different height (3 meters-30 meters), the shrubs (below 3 meters) and the herbs. Relatively good rainfall, high humidity, optimum temperature and presence of sufficient space condition the development of forests.

Atmosphere of Forests

The forest environment is less variable. Its uppermost level (tree tops) forms a canopy, which itself is exposed to sunlight and wind fury, but it does not permit them to the ground. As a result:

(i) Only diffused light reaches the ground.

(ii) Moisture contents of the soil remain more or less stable and are maintained for longer periods because of absence of evaporation.

(iii) There is no loss of water by transpiration below the forest canopy, because the tree tops act as effective wind breakers.

(iv) The temperature variations are peculiar. Temperature is generally lower in summer and is higher in winter. Also the days are cooler and nights are warmer than the surroundings.

(v) The floor of forest is thickly carpeted with encrusting thalloid plants like lichens and mosses. Several herbs also grow at the ground level.

Kinds of Forests

Great variations have been observed in the nature and abundance or sparseness of forest vegetation. On that basis following types of forest biomes have been recognized.

1. Northern coniferous forests
2. Temperature forests
3. Tropical forests
4. Subtropical forests.

Northern Coniferous Forests: Coniferous forests are met with in the Northern Hemisphere, restricted to northern areas of North America, Europe, parts of Japan and Siberia. In Southern Hemisphere it is found in Australia and southern parts of South America. In Northern Hemisphere, its extensive geographical belt south of Tundra is known as 'Taiga'.

Climate: Coniferous forests are found in cold regions with heavy rain falls, high humidity and strongly seasonal climate. It has long winter with fairly short summer of about 60-150 days mean monthly temperature varies between –30°C (winter) and +23°C (summer) and faces heavy snow fall. The soils are acidic and mineral deficient and poor in nutrient. Because of the absence of evaporation, the soluble essential nutrients (calcium, nitrogen and potassium) are leached down sometimes even beyond the reach of roots.

Vegetation: Abounding these forests are coniferous evergreen trees of pine, balsam, larch, spruce and fir. These are the chief source of timber. The understorey trees are uncommon because of low light penetration. Orchids and ericaceous shrubs like blue berry constitute the only understorey vegetation. Mosses and lichens are found in abundance.

Fauna: The vertebrate fauna is represented by mammals like lynx, bears, wolves, red foxes, squirrels, porcupines, star-nosed moles, lemmings, mouse, hares, deers, several groups of large sized mammals like moose, elk, caribou, beavers and wolverine etc. The common birds of this region are grouse, cross-bills, jayas, woodpeckers, flycatchers, saw-whet owl, grosbeak, sparrow, raven, wren, thrushes, warblers, finches and pigeon-hawk etc. Hylids are the only amphibians of coniferous forest and reptiles are few.

In the foot hills of the forest zone of India are found the tiger, the elephant, the Sambar deer, the gaur, the chital and the swamp deer The invertebrate fauna is dominated by insects like wood-wasps, bark-bettles, pine sawflies and pine-geometrids. Snails, annelids and millipedes are scarce.

Role of Temperate Forests

The temperate forests are characterised by moderate climate with seasonal variations. These are of three types:

(1) *Temperate Forests:* These are met within North America, Western Europe, Eastern Asia (China and Japan), Chile and parts of Australia. The temperate deciduous forests have hot summer (maximum temp. 20-27°) and cold winter (–12°C) with snowfall. The rain comes in summer and the average rainfall varies from 30-50 inches per year. The summer, autumn, winter, spring and rainy seasons are distinctly merged. These seasonal changes have greatly effected the vegetation.

The deciduous forests are characterised by broad leaf trees, which attain a moderate height of 40-50 meters and form a thick canopy during summer. But during winter, the leaves are shed to withstand the excessive cold and drought. The principal trees are pines, oaks, hickory, maple, cidar, cotton wood, basswood, beech, chestnut, elms and willows. The understorey of shrubs and herbs is well developed, but herbs are more numerous. Often there are two blooms of herbs, the spring flower bloom immediately after winter before the trees have expanded their leaves, and the rainy season bloom which grows under low light level. There are a few vines (the grape) and some epiphytes, mosses, algae and lichens growing on tree trunks.

The invertebrates are represented by the earthworms, millipedes, snails, land isopods, wood-boring larvae of various insects and the leaf eating insects. The mammals are represented by squirrels, deer, bears, gray foxes, bobcats, pigs, dormice and ground dwelling wild boar, wolves, racoons, badger and shrew, goat, antelope, wild cat, lynx etc. The flying mammals include bats, flying squirrels, arboreal porcupines and opossums. The avian fauna is represented by hawks, owls, woodpeckers, thrushes, crested flycatcher, horned owl and a good variety of crows, ravens, herons and storks, etc. The reptiles are represented by turtles, timber rattle snakes and amphibians by newts, salamanders, toads, tree frogs and cricket frogs etc.

The deciduous forests are dominated by ground dwelling and cursorial forms.

(2) *Temperate Evergreen Forest:* This forest is characterised by warm dry summers and cool, moist winters. It is represented in North America (chaparrel of Pacific Coast), maquis in Mediterranean region, encinar in Spain and melle scrab on Australia's south coast.

In evergreen woodland the trees are absent, but shrubs range upto 3-4 meters in height. The forms are intermediate between the temperate deciduous forest and the drier grassland biomes. The plants possess great power of regeneration and sprout very quickly after being burnt by forest fire.

Animals characteristic of this woodland are wood rats, chipmunks, brush rabbits, mule, deer, lizards, wem-tits and brown towhees. The mammals are mostly cursorial ungulates, and jumping or saltatory forms.

(3) *Temperate Rain Forest:* The temperate rain-forest has a definite seasonality. The temperature as well as rainfall vary throughout the year. Rainfall is high but often fog is very thick and serves as an important source of water. Therefore, the temperate rain forest is the coldest.

Due to availability of water throughout the year, the diversity in flora and fauna is less. The dominant trees are redwood of Pacific Coast of North America; alpine ash of Australia and Tasmania. These are very high measuring even more than 100 meters in height. Epiphytes and lianas are common but not so rich as in any other rain-forests. Fauna is similar to other temperate forests.

Role of Tropical Forests

Tropical Forests are found in tropical zone of the world and are characterised by very high temperature with rainfall in abundance. These are of two types:

(i) *Tropical Rainforests:* Tropical rain forests occur near the equator in Central and South America, Central and Western Africa, Southeast Asia (Malaya and parts of India, New Guinea, Java, Sumatra, Malaya, Borneo) and Northeast Australia. These forests are characterised by minimal seasonal changes. These have high and even temperature throughout the year. The maximum temperature ranges between 32°C-40°C and minimum between 10°C-0°C. The days and nights are of uniform length. The annual rainfall varies from 250-400 cm (100-160 inches) and generally it is uniformly distributed throughout the year. Due to high temperature, there is speedy evaporation of water from the heated land surfaces. When these vapours rise up these are cooled and result in heavy down-pour accompanied by lightening and thunder mostly in the afternoon. The relative humidity is always very high and seldom drops below 60 percent of saturation.

Soils in the tropical rain forests are red latosols and are exceedingly thick. Leaching of minerals at a very fast speed makes these soils unfit for agriculture. But there is speedy cycling of nutrients which makes these soils very fertile.

The flora of tropical rain forest is very rich and highly diversified. One square mile of land may contain 300 different species of trees. The trees are very high *i.e.* 36 to 56 meters (100-180 fit) and rich in thick stemmed woody lianas.

The tropical forests have a very rich fauna both in density as well as in varieties. The invertebrate and cold blooded vertebrates have attained the largest possible size. The common invertebrates are planarians, worms,snails, millipedes, centipedes, scorpions, spiders, isopods and insects. Among insects bugs, beetles, ants, bees, flies, termites, mantids, phasmids and some other orthopterans like basalattids are very numerous. Land leeches are found on the floor of forest as well as on the bushes.

The vertebrate fauna is very different in its different subregions and is dominated by arboreal and nocturnal forms. The common mammals found in forests are tiger, panther, leopard, lion, jackal, elephant, deers, sambar deer, chital deer, bison, tapirs, hippopotamus, scaly anteaters, wild pig, jungle cat, giant flying squirrel, flying phalangers, monkeys, howling monkey, spider monkey, black langur, gorillas.

The birds are represented by a large number of varieties and display vivid colours. These are woodpeckers, fruit eaters, humming birds, parrots, hornbills etc. The reptiles of tropical forest are geckos, chameleons, agamid lizards and many species of poisonous and nonpoisonous snakes and some aquatic reptiles. Frogs and toads are very common and arboreal amphibians are represented by Rhacophorus.

(ii) *Tropical Seasonal (=deciduous) Forests:* These are found in south-east Asia (India), Central and South America, Northern Australia and the tropical island of the Pacific and Indian oceans. The tropical seasonal forests are characterised by seasonal variations. The temperature and rains both are quite high, but the rainfall is not uniform throughout the year. The periodic rains (monsoons) result from the outflow of dry and hot winds from high pressure areas and an inflow of moisture-laden winds from the surrounding oceans.

The seasonal or deciduous forests are also known as monsoon forests. These are less lofty than the evergreen tropical forests and are almost leafless during dry season. The trees are tall well over 40 meters in height, but more commonly these are about 20-30 meters high.

These are arranged in two strata. In the upper strata, the canopy is continuous and deciduous, but some trees in the lower strata are evergreen. These forests are rich in woody lianas and herbaceous epiphytes. Teak is the common tree and bamboo is the most common shrub.

The animal fauna is similar to tropical rain forests except that it is slightly less varied.

Rainforests of Subtropicals

The subtropical region of Florida in America has fairly high rain and is characterised by broad-leaved evergreen vegetation. The vegetation includes mahogany, gumbo, limbo bays, palms, oaks, mangolias and tamarind etc. All these are associated with epiphytes, ferns and vines. Animal life of subtropical forests is similar to rainforests.

Animal Adjustments: The forest animals are comparatively strong and muscular and most of them are adapted to arboreal habit (for details see arboreal adaptations). These have a better sense of hearing and exhibit protective colouration. Their breeding habits are arboreal.

Biomes of Grassland

The grassland biomes are found both in the middle lattitude and equatorial tropical climatic zones. The major controlling factor is the rather sparse erratic rainfall of about 10-40 inches annually. The grasses vary from short buffallo grass to tall elephant grass and bamboo.

There are various regional designations for the grassland biomes:

(1) *The Savannas:* These are found in the tropical plains of Africa, India and South America. The climate is characterized by a long dry winter season alternating with a wet summer season. The temperature remains relatively warm throughout the year.

(2) *The Prairies:* These are widespread in Central U.S.A., Eastern Europe and in the South American countries of Argentina and Uruguay. There are great annual variations in temperature because of their middle

lattitude location, and rainfall tends to be scanty. The grasses vary from tall to short species.

(3) *The Steppe:* The steppe is drier than the prairie, contains short grass species and is located in interior plains and plateaues that are semiarid. Steppe grasslands are found in Western North America, but are most extensive in Central and West Asia.

***Fauna*:** The inhabitants of grassland are antelopes, hippopotamus, horses, asses, dogs, wild beasts, ground squirrels, cattle which graze upon grass, frogs, limbless amphibians, reptiles including turtles, tortoises, lizards and snakes and the grain eating birds. Insects, grass-hoppers, termites, ants, bees and wasps are dominant varieties.

The inhabitants of grassland are characterized by the power of swift running, flocking habit and keen sense of sight.

Biomes of Desert

Desert is that arid region of the earth surface, where rainfall and moisture are insufficient to support normal life. In these regions average rainfall never exceeds 10-15 inches per year. The rains are uncertain and may not occur even for years together. Depending upon the rains, deserts can be categorised into:

(1) low rainfall deserts *e.g.* Atacama,

(2) cold deserts such as Alps and Scandinavian mountains,

(3) hot deserts such as found in equatorial region,

(4) low nutrient deserts such as North American deserts,

(5) deserts due to toxic conditions as found near volcanoes and

(6) high salt deserts found in Chile, South America and Australia.

World Deserts

Deserts occupy about one-seventh of the land surface of the earth and are largely confined to world wide belt between Tropic of Cancer and the Tropic of Capricon. Deserts are found in almost all the countries of the world. Some of them are widely expanded, whereas others have a short range.

1. The Great North American desert covers Western United States, North Mexico, Central Mexico, South East California, South West Texas, Arizona and Nevada.
2. South American deserts are found in Atacama and Argentina. The former extends along the coast of Peru and Northern Chile and the later includes semi-arid regions of Andes.
3. In Europe the deserts are restricted to Alps and Scandinavian mountains.
4. The Asian deserts are Thar desert in North-West India (Rajasthan), Indus Valley in Pakistan and Gobi desert in China.
5. The African deserts are Sahara in Northern part and Kalahari in Southern region.
6. In Australia the central and western parts of the continent have desert conditions.

Environment of Desert

Desert climate is characterized by following features:

1. *Scarcity of Water:* Scarcity of water or dryness is characteristic of deserts. This is due to rainfall and absence of natural sources of water.
2. *Extremes of Temperature:* The temperature in day time is very high and relative humidity is very low. The scorching sunlight falls directly on living organisms which becomes unbearable. During night the

temperature goes down tremendously because of the radiation of heat into space. Thus deserts present both the extremes of temperature within one day.

3. *Dust Storms:* Lack of moisture and streams of temperature cause an increase in the air movement and dust storms almost every evening.
4. *Lack of Vegetation:* Due to scarcity of water, plants are unable to grow. These occur only around areas, where very little water is available. The vegetation is also modified for this mode of life being succulent and thorny, which prevents excessive evaporation of water and helps in its conservation.

Vegetation of Desert

The aridity of desert severely limits plant growth. The vegetation is very thin and is mainly found near oases. The plants which grow in desert are either drought-evading (ephemerals) or drought-resisting.

(i) The drought evading plants or ephemerals lie dormant as seeds and sprout only when moisture and temperature are favourable. Such plants have brief life-cycle to be completed within weeks. The African Boerhavia repens produces seeds even after eight or ten days of sprouting.

(ii) The drought resisting plants have evolved several means of storing water, locating underground water or reducing evaporation of water. Typical drought resisting plants are succulents which have very small or even no leaves. The steam is succulent and stores water. Their surface is covered by thick waterproof skin and the leaf-pores can be tightly shut. Some of them store water in tuberous roots. Most common examples of desert succulents are cacti and euphorbias, whose fleshy tissues hold vast amounts of liquid, and whose shallow, widespread roots absorb surface moisture.

(iii) Another variety of drought resistors are the shrubs such as creosote bushes, acacias, cassias and sage bushes etc. They possess short basal trunk with branches having short and thick leaves and thorns.

Animals of Desert

Under the harsh and shimmering sun the arid baked sand of the desert may appear to be devoid of animal life, but surprisingly large number of animal species thrive there. Like plants, these are also either drought evaders or drought resistors. Like ephemeral plants, drought evader animals also make their appearance only when rain sets in, for the rest of the time they remain dormant either in egg or pupal stage. Insects like crickets, grasshoppers, ants, bees, wasps, butterflies, moths and beetles swarm in the desert, when rains arrive and plants flourish. Tiny shrimps appear in temporary ponds, and lizards and snakes are also very common. Birds are few and live in places which have surface water for drinking within their flying range. Small rodents such as kangaroo rat, pocket mice and desert mangoose, desert fox, desert hedgehog, kit-fox, Jace-rabbit and the camels, deers, antelopes and asses are the mammalian forms found in deserts.

Desert-like Adaptation

The fauna and flora of desert is peculiarly adapted to these drastic conditions and the modifications are mostly associated with the conservation and availability of water, protection against extreme heat and cold and obtaining food.

1. *Obtaining Water:* Higher animals cannot live long without drinking water and need water at frequent intervals. But scarcite of water is the primary cause for desert conditions. Free water is not available for miles

together. Therefore, desert animals try to utilize water from every available source.

(i) In most deserts water holes or springs are present as in Australia. Surface dwelling animals make use of this water and conserve it for long. If this water is not available these animals are unable to survive.

(ii) A few animals absorb dew drops along with plants food.

(iii) Certain desert animals such as desert rabbit, turtles and wood rat (Neotoma) do not drink water even if it is available to them. These suffice their need by eating succulent plans.

(iv) Still another group or desert animals like Jerbous, kangaroo rat (Dipodomys), pocket mouse (Perognathus) and other desert rodents can live on the water received from eating dry seeds and vegetation.

(v) Carnivorus animals get needful water from the body of their prey. The grasshopper-mice of western North America, long eared hedgehog of Sahara and mulgara, a marsupial which inhabits the most arid zone eat insects and small rodents.

(vi) The most specialized group of animals (ant and rats-Jerboas and kangaroo rat) depends entirely upon metabolic water produced in their body during the oxidation of foodstuffs.

The oxidation of fat releases maximum water while minimum amount is released in the oxidation of proteins.

1 *g* of fat yields 1.1 gram water,

1 *g* carbohydrate yields 0.6 gram water, and

1 *g* protein yields 0.3 gram water.

(vii) Skin of certain animals like Moloch is hygroscopic and absorbs water like a blotting paper.

(viii) The young ones of several rodents are born at a time that coincides with the short rainy season so that the fluid lost through milk production is compensated by juicy grass.

2. *Water Conservation:* Terrestrial animals may lose water from their body in three ways:

(a) By evaporation through the body surface;

(b) By exhaling moisture during respiration; and

(c) By expelling water or liquid urine during excretion. Animals living in desert have developed means to avoid water loss by all these ways.

(a) *To Avoid Evaporation Through Body Surface:* Desert animals have developed several protective measures to avoid loss of water by evaporating through general body surface:

(i) Desert animals develop thick hide which avoids water loss by perspiration.

(ii) Number of sweat glands in the skin of desert mammals is very much reduced or in some cases these are totally absent.

(iii) Scorpions and reptiles have developed almost impermeable outer covering. Scales and spines are developed on the exposed surfaces. Examples are sand lizard (Moloch) and horn toad (Phymosoma).

(iv) Some desert animals produce epiphragms and secretions to prevent loss of water. Desert insects are wax-proof.

(v) Certain animals like camel have developed more tolerance to heat. Their body temperature fluctuates more with the atmosphere and thus reduces the amount of moisture lost through sweating and panting. Camel can lose 30 per cent of its body weight without showing signs of distress and protein deficiency. Mostly animals loose water from the blood which becomes thick or dehydrated causing death. In camel water is lost from the tissues and not from the blood.

(b) *To Avoid Loss of Moisture During Respiration:* Many desert mammals cool the exhaled air in their nose, before it is expelled out through nostrils. As a result moisture condenses in the nose and is not lost as water vapours along the exhaled air.

(c) *To Avoid Loss of Water During Excretion:* This is brought about by excreting concentrated urine or urine in semisolid state. Desert rats and kangaroo rats excrete urine which contains 24% urea, whereas in man urine contains just 6% urea. Even reptiles, insects and birds pass their body wastes in such concentrated state as uric acid that little or no moisture is lost in the process. Uric acid can be crystallised in the body. In camel also the amount of urea in the urine increases and the amount of urine decreases with the reduction in the amount of water conserved in the body.

(d) Other methods of conserving water:

(i) Desert animals are mostly burrowing and nocturnal. They come out of their burrows either late in night or early in the morning, when percentage of moisture in air is almost equal to the moisture in their burrow. The burrow protects them against scorching sunlight and avoids excessive water loss due to evaporation. SCHMDIT and NIELSEN'S have found that relative humidity in the burrows was 30-50% whereas outside the burrow in the air it was 0-15%. Certain animals are found to plug the mouths of their burrows during day time.

(ii) Desert animals are active only far a brief period during those part of the year when water and food are available. For rest of the time they remain dormant lying hidden in some burrows or holes.

(iii) Uromastix is said to store water in large intestine and in camel water is stored in the rumen along with fermenting food in a pocket-like structure. But the recent investigation suggest that the liquid present in rumen is not actually water but its composition is similar to digestive juices and its amount is far less than present in the stomach of other animals.

3. *Protection Against Scorching Sun:* Desert animals have to protect themselves against excessive heat. The temperature of the air may boost up upto 45°C and at that time the sand is still at a higher temperature and may be at 57-58°C. For protection against scorching sun and burning sand, desert animals have developed protective armours and several defensive mechanisms:

(i) Desert animals have thick hide or have a protective armour of scales, spines or dermal scutes.

(ii) Desert animals have long legs, which no doubt help in jumping and swift running, but also lift the body above the ground and thus avoid direct contact of the body with burning sand, *e.g.* leaf hoppers, kangaroo rat, and Jerboa etc.

(iii) The simplest and most wide-spread method of escaping desert heat is by burrowing. The burrows are humid and cooler. The air in the burrow acts as insulation. The desert animals hide themselves in the burrows during day time and come out only late in evening at night or early in the morning when temperature is low and relative humidity is more in the air.

(iv) Many desert animals have physical adaptations to enable them to burrow and survive under the sand. Scorpions, some lizards, certain beetles and desert mongoose are the burrowing animals. Scorpions have enlarged claws while lizards have wedge-shaped heads which cleave the sand. Certain species of beetles become so flattened that they can burrow in the sand with side to side movement of the body. Some lizards can dive in the sand with head first as in water.

(v) Some desert animals are acclimatized to the high temperature of desert and to restore their body temperature they bask in the sun. Beetles and lizards in general could endure temperature so high which cannot be tolerated even by human beings.

(vi) Other animals do not burrow but hide themselves under some suitable shelter and avoid the scorching sunrays of the noon.

4. *Protection Against Sand Storms:* The desert are often found to have dust storms in the afternoon and these are sometimes so strong that heaps of sand are washed away very swiftly. The important organs affected by the drifting sand are the nostrils, ears and eyes. Several mechanisms have been evolved to protect these structures:

(i) Nostrils in majority of desert animals are either reduced to small pinholes or they are protected with complicated valves. For example, in camel, the nostrils can be closed in the same way as eyes are closed by the eyelids. In sand-swimming lizards the nostrils are turned upwards and are thus protected from sand.

(ii) Eyelids exhibit various modifications and are capable of closing the eyes without hindering the vision. In Mabuya the lower eyelid is much enlarged and transparent and can be drawn on the eye like the nictitating membrane. It spreads over the complete eye thus protecting it against the sand but permitting vision. In certain other lizards (Ablepharus) the lower eyelid is permanently stretched over the eye and fused with the lower margin of upper eyelid. It is wholly transparent. Thus their eyes are permanently closed but are able to perceive vision. In camel the eyes are high on the head and are protected by long and thick eye lashes.

(iii) Ears are often protected with fringes of scales of hairs. In Phrynocephalus and Phrynosoma the air apertures are absent. Nocturnal animals need a sharp sense of hearing. Jack rabbits, hares etc. have large ears.

5. *Swift Running or Speed:* Since desert animals have to move far and wide in search of food and water, majority of them have great speed. In addition, for protection against enemies also these have to rely on speed. The limbs are specialized and adapted to walk on sand.

(i) The limbs are often long in desert animals, for example, leaf hopper and kangaroo rat. The latter has owned its name only on account of specially long hindlimbs.

(ii) Desert cat has wide soles thickly covered with fur which enables it to walk comfortably even on hot sand.

(iii) Legs of ostrich are padded with heavy callosities.

(iv) The undersurface of camel's feet is also padded.

(v) In Phrynocephalus and Teratonrhycus the digits possess lateral fringes.

(vi) In Scapteria the digits are broadened out into shovels.

(vii) American collar lizard (Crotaphytus collaris) and gridiron tailed lizard (Callisaurus) have longer hindlimbs and are capable of, running very fast. The latter is known to move on hindlimbs only with tail acting as a balancing organ.

6. *Colour:* The colouration in desert animals is found to match with the sand dunes. The body hues are often grey, brown or red, which is often in harmony with the ground. Often the colour is dull which blends with the surroundings. This blending colouration furnishes protection from the attack of predators by camouflaging. In the absence of definite places for concealment the colour-blending is the only measure of protection and the animals cannot be easily detected in moonlit nights. Warning colouration is exhibited by poisonous animals like lizard, Heloderma, rattle snake, spiders and red ants. In general, desert animals are less heavily pigmented and are comparatively small in size (Gloger's rule).

7. *Spinescence:* The spine studded body is characteristic feature of desert animals. The most conspicuous examples are Moloch horridus (flat spiny lizard), Phrynocephalus (horn toad) and Phrynosoma (horn lizard).

8. *Venom:* Possession of venom is another attribute to desert animals for self protection. Helodenna punctatum is the only poisonous lizard and is found in the arid zone of Mexico. Rattle snakes, trap door spider and Tarantula spiders are poisonous creatures of the desert.

9. *Sense Organs and Intelligentia:* Organs of sight, smell and hearing are specially developed. The jack rabbits and Jerboas have long external ears which collect and concentrate the sound waves from the air. These are able to see in the darkness of night. In desert mammals the tympanic bulla is enlarged. This probably adds greatly to the sensitivity of the animal's hearing. Certain desert animals exhibit very intelligent behaviour. Puff lizard when attacked, darts into a crevice between two rocks and inflates the body to such an extent, that it cannot be pulled out. Kangaroo rat when attacked by a rattle, throws sand and gravel in the face of the snake with the help of its hind-legs.

Reptiles and some insects appear to be preadapted to desert ecotype. Many reptiles have structural and behavioural adaptations to cope with sandy habit. Mammals as a group (except a few) are poorly adapted to desert life. But as such the desert fauna exhibits a very good example of adaptation to the environment.

Biome of Mountains

Mountains often compress many different biomes of life zones into a small area because of the differences in altitude. In Southern Mexico we can find almost all the ecosystems or biomes. At the lower altitudes, there are tropical rain forests. As one ascends, it is noted that at the eastern slope of the mountains the rain forest blends into the typical deciduous forest with a climate similar to that of deciduous biome.

Continuing up the mountains is the coniferous forest with a climate similar to that of Southern Canada (coniferous forest zone). Finally is the area of perpetual ice and snow at the peaks of high mountains, which is comparable to polar regions of Tundra. The western slopes of mountains have desert condition at lower altitudes.

Animal life accords with the climatic conditions. Even the plant fauna is also similar. But there is some difference in the dispersion of animals. Except birds which have ability to fly, the other vertebrates do not reach the top of the mountains.

The land biomes are not always marked out sharply, but they exhibit overlappings and merging and they do not follow the lines of latitudes because of variation of altitude and variations in climate.

Biome of Caves

Caves are natural hollows under the earth surface or in mountains or hills. These actually represent abandoned channels through which underground rivers must have flowed in the past. Majority of caves are formed by the dissolution of limestone and are found only in those regions where massive limestones are deposited near the earth surface. Since the limestone is soluble in water, the surface water gradually sinks underground forming crevices, which increase in size sideways and form the channels, known as caves or caverns.

Caves of the World: Caves are found all over the world but are restricted to particular areas. The most famous cave area is located in the calcareous Alps of central and southern Europe extending through France to Austria and Yugoslavia. In America it extends through parts of Indiana. Kentucky, Tennessee and Missori. Well known caves are present in Himalayas, Malaysia, Indonesia, South Africa and Australia. Dry caves are present in the rocks of Gibralter, Malta and

Lebanon. Caves are also found in Europe especially in Carniola, Austria and Illyria. Basses, Pyreneest in France and Spain, and Dordogne region in Western France and others are well known caves in which Palaeolithic man had lived.

Environment of Caves

The cave environment presents stable conditions for living beings. The environmental conditions are more or less similar in different caves.

Caves are characterised by the absence of bright light and nearly uniform temperature, high humidity, feeble fluctuations of temperature and absence of air currents. But light is present at the mouth of the caves which penetrates for a short distance inside. Therefore, the cave environment is separated by LULL into three distinct regions, namely:

(i) Twilight or dysphotic or transitional region.

(ii) Region of fluctuating temperature.

(iii) The inner cave region.

The latter represents true cavern environment having no light, relatively high humidity and absence of air currents. The absence of light has a pronounced effect on the inhabitants of this region.

Cave Flora: The absence of light does not favour life and growth of green plants which are used as food. Therefore, only those plants exist which can bear the darkness. These are Ozonium (common fungus), green mould, long white mould, mushrooms (Agaricus sp.).

Cave Fauna: Cave dwellers are classified into three groups:

1. *Permanent Cave Dwellers:* The permanent cave dwellers reside permanently in caves and do not come out of them. These include both terrestrial as well as aquatic forms and are represented from almost all the phyla:

turbellarians from Platyhelminthes; chaetopod worms and leeches from Annelida; amphipods, isopods, decapods etc. from class Crustacea; springtails from class Insecta; mites and true spiders from Arachnida and snails from Gastropoda. The cave dweller vertebrates are represented by all the classes. Amongst fishes are Amblyopsis. Syphlichthyes chologaster and the cave catfish (Gronias nigrilabris). In addition, large number of fishes of the family Amblyopsidae are found in North American caves. Cave amphibians are cave salamanders such as Spelerpes maculicauda, Typhlotriton spelaeus, Proteus anguinus. Nocturnal owls and oil birds nest in the caves but are not permanently cave dwellers. Amongst mammals white footed mouse (Peromyscus leucopus) is the only cave dwelling form.

2. *Occasional Cave-dwellers:* These constitute the temporary fauna of caves and include bats and owls.

3. Accidental cave dwellers or strugglers.

Cave Adaptations: The body organization of permanent cave dwellers has undergone marked changes, but these are mostly towards degeneration. The obvious reason is that cave environment presents simplified ecological conditions, characterised by uniform and rather unchangeable atmosphere, lack of competition, absence of natural enemies. Therefore, cave animals are known for their primitive organization and defenceless nature.

1. *Loss of Pigmentation:* Cave animals are either colourless or pale white in colour. The loss of pigmentation is on account of absence of light in the caves. If coloured, they are lighter or paler than their epigean relatives. Moreover, certain cave dwellers when exposed to light, develop pigments. For example, Niphargus (a cave dweller amphipod) when reared in light, develops

pigment in a short period. Similarly, Proteus develops dark pigment when exposed to light. In certain cases the light is detrimental. Spring tails and mites may die when exposed to light.

2. *Slender Body and Reduced Appendages:* The cave dwelling forms are often found to possess elongated and slender body with attenuated appendages. For example, in crustaceans, Bathynella and Parabathynella, the antennules and antennae are slender. In urodeles the body is slender and limbs are short and slender.
3. *Weak Exoskeleton:* The cave arthropods have weak or fragile exoskeleton. The shell is thin in cave snail. In cave fishes scales are lost altogether. The insects dwelling in caves are thinner than their surface dwelling relatives.
4. *Absence of Visual Organs:* Since light is absent in caves the eyes are either reduced, degenerated or absent in cave animals, or these are embedded in the integument. The eyes are completely absent in cave fishes *e.g.* Stygicola and Leucifuga from Cuba and Amblyopsis from mammoth caves of America. Among amphibians eyes are nonfunctional in Typhlomolge and Typhlotriton and in Apodon and Ichthyophis these are covered by a fold of integument and appear as mere black dots.
5. *Tactile Organs:* To compensate the impaired sense of sight, other senses are highly developed in cave animals. The cave dwellers often depend either on smell or on touch. In cave insects, myriapods and arachnids the legs and antennae are long and provided with tactile hairs. The head, snout and lips of cave fish Amblyopsis bear taste organs in the form of terminal buds. Chemoreceptors are well developed in cave animals

than in their epigean relatives. For example, cave isopod, Asellus has large sense bulbs.

6. *Retrogression of Endocrine System:* In cave animals the endocrine system exhibits varying degree of degeneration, especially in cave urodeles. The thyroid gland in Proteus is said to secrete a hormone which has no effect on the body tissues.

In addition to the above modifications the cave animals are characterised by low metabolic rate, very low fecundity and slow rate of development. There is no competition or struggle for existence among cave dwellers. Hence instead of specialization, these have under gone degeneration. Further the environment is roughly constant with no major changes, even for hundreds of years together. These are lethargic creatures.

Therefore, archaic forms have been able to establish themselves. The snails of the genus Lartetia, Bathynella, Proteus (salamander), Troglochaetus (annelid), and Cirolanides (an isopod) are reptiles of the past. The crustacean Bathynella belongs to palaeozoic group, Syncardia.

Cave dwellers possess such primitive features as presence of lateral line system on the head of Caecobarbus and the ossification of hyoid and branchial arches in Proteus (an amphibian). Neotenic forms are also found in cave dwellers *e.g.* Proteus and Siren.

The Biodiversity

If we divide the whole earth's mass into 10 billion parts, it is only in one part where life exists and the astounding variety of living organisms numbering somewhere around 50 million species are all restricted to just about a kilometer-thick layer of soil, water and air. Isn't it wonderful to see that so much diversity has been created by nature on this earth from so little physical matter! Biodiversity refers to the variety and variability among all groups of living organisms and the ecosystem complexes in which they occur.

From the driest deserts to the dense tropical rainforests and from the high snow-clad mountain peaks to the deepest of ocean trenches, life occurs in a marvellous spectrum of forms, size, colour and shape, each with unique ecological inter-relationships. Just imagine how monotonous and dull the world would have been had there been only a few species of living organisms that could be counted on fingertips!

In the Convention of Biological diversity (1992) biodiversity has been defined as the variability among living organisms from all sources including inter alia, terrestrial, marine and other aquatic ecosystems and the ecological complexes of which they are a part.

Biodiversity and its Phases

Units of biodiversity may range from the genetic level within a species to the biota in a specific region and may extend up to the great diversity found in different biomes.

Reasons of Diversity

It is the basic source of biodiversity. The genes found in organisms can form enormous number of combinations each of which gives rise to some variability. Genes are the basic units of hereditary information transmitted from one generation to other. When the genes within the same species show different versions due to new combinations, it is called genetic variability. For example, all rice varieties belong to the species *Oryza sativa*, but there are thousands of wild and cultivated varieties of rice which show variations at the genetic level and differ in their colour, size, shape, aroma and nutrient content of the grain. This is the genetic diversity of rice.

Different Kinds of Species

This is the variability found within the population of a species or between different species of a community. It represents broadly the species richness and their abundance in a community. There are two popular indices of measuring species diversity known as Shannon- Wiener index and Simpson index. What is the number of species on this biosphere?

The estimates of actual number vary widely due to incomplete and indirect data. The current estimates given by Wilson in 1992 put the total number of living species in a range of 10 million to 50 million. Till now only about 1.5 million living and 300,000 fossil species have been actually described and given scientific names. It is quite likely that a large fraction of these species may become extinct even before they are discovered and enlisted.

Diversity in Ecosystem: This is the diversity of ecological complexity showing variations in ecological niches, trophic structure, food-webs, nutrient cycling, etc. The ecosystems also show variations with respect to physical parameters like moisture, temperature, altitude, precipitation, etc. Thus, there occurs tremendous diversity within the ecosystems, along these gradients. We may consider diversity in forest ecosystem, which is supposed to have mainly a dominance of trees. But, while considering a tropical rainforest, a tropical deciduous forest, a temperate deciduous forest and a boreal forest, the variations observed are just too many and they are mainly due to variations in the above-mentioned physical factors.

The ecosystem diversity is of great value that must be kept intact. This diversity has developed over millions of years of evolution. If we destroy this diversity, it would disrupt the ecological balance. We cannot even replace the diversity of one ecosystem by that of another. Coniferous trees of boreal forests cannot take up the function of the trees of tropical deciduous forest lands and vice versa, because ecosystem diversity has evolved with respect to the prevailing environmental conditions with well-regulated ecological balance.

Classification of India on Biogeographical Grounds: India has different types of climate and topography in different parts of the country and these variations have induced enormous variability in flora and fauna. India has a rich heritage of biological diversity and occupies the tenth position among the plant rich nations of the world.

It is very important to study the distribution, evolution, dispersal and environmental relationship of plants and animals in time and space. Biogeography comprising of phytogeography and zoogeography deals with these aspects of plants and animals. In order to gain insight about the distribution and environmental interactions of flora and fauna of our country, it has been classified into ten biogeographic zones as given below. Each of these zones has its own characteristic climate, soil, topography and biodiversity.

India's Major Biogeographic Habitats

Sr. No.	*Biogeographic Zone*	*Biotic Province*	*Total area (Sq. Km.)*
1.	Trans-Himalayan	Upper Regions	186200
2.	Himalayan	North-West Himalayas	6900
		West Himalayas	720000
		Central Himalayas	123000
		East Himalayas	83000
3.	Desert	Kutch	45000
		Thar	180000
		Laddakh	NA
4.	Semi-Arid	Central India	107600
		Gujarat-Rajwara	400400
5.	Western Ghats	Malabar Coast	59700
		Western Ghat Mountains	99300

Contd...

Sr. No.	*Biogeographic Zone*	*Biotic Province*	*Total area (Sq. Km.)*
6.	Deccan Peninsula	Deccan Plateau South	378000
		Central Plateau	341000
		Eastern Plateau	198000
		Chhota Nagpur	217000
		Central Highlands	287000
7.	Gangetic Plain	Upper Gangetic Plain	206400
		Lower Gangetic Plain	153000
8.	North-East India	Brahmaputra Valley	65200
		North-Eastern Hills	106200
9.	Islands	Andaman Islands	6397
		Nicobar Islands	1930
		Lakshadweep Islands	180
10.	Coasts	West Coast	6500
		East Coast	6500

Source: "Conserving our Biological Wealth", WWF for Nature-India and Zoological Survey of India.

Biodiversity's Value: The value of biodiversity in terms of its commercial utility, ecological services, social and aesthetic value is enormous. We get benefits from other organisms in innumerable ways. Sometimes we realize and appreciate the value of the organism only after it is lost from this earth. Very small, insignificant, useless looking organism may play a crucial role in the ecological balance of the ecosystem or may be a potential source of some invaluable drug for dreaded diseases like cancer or AIDS. The multiple uses of biodiversity or biodiversity value has been classified by McNeely *et al* in 1990 as follows:

(i) *Consumptive use Value:* These are direct use values where the biodiversity product can be harvested and consumed directly *e.g.* fuel, food, drugs, fibre, etc.

Food: A large number of wild plants are consumed by human beings as food. About 80,000 edible plant species have been reported from wild. About 90% of present-day food crops have been domesticated from wild tropical plants. Even now our agricultural scientists make use of the existing wild species of plants that are closely related to our crop plants for developing new hardy strains. Wild relatives usually possess better tolerance and hardiness. A large number of wild animals are also our sources of food.

Drugs and Medicines: About 75% of the world's population depends upon plants or plant extracts for medicines. The wonder drug Penicillin used as an antibiotic is derived from a fungus called Penicillium. Likewise, we get Tetracycline from a bacterium. Quinine, the cure for malaria is obtained from the bark of Cinchona tree, while Digitalin is obtained from foxglove (Digitalis) which is an effective cure for heart ailments. Recently vinblastin and vincristine, two anticancer drugs, have been obtained from Periwinkle (Catharanthus) plant, which possesses anticancer alkaloids. A large number of marine animals are supposed to possess anti-cancer properties which are yet to be explored systematically.

Fuel: Our forests have been used since ages for fuel wood. The fossil fuels coal, petroleum and natural gas are also products of fossilized biodiversity. Firewood collected by individuals are not normally marketed, but are directly consumed by tribals and local villagers, hence falls under consumptive value.

(ii) *Productive use Values:* These are the commercially usable values where the product is marketed and sold. It may include lumber or wild gene resources that can be traded for use by scientists for introducing desirable traits in the crops and domesticated animals. These may include the animal products like tusks of elephants, musk from musk deer, silk from silk-worm, wool from sheep, fir of many animals, lac from lac insects, etc. all of which are traded in the market. Many industries are dependent upon the productive use values of biodiversity *e.g.*- the paper and pulp industry, plywood industry, railway sleeper industry, silk industry, textile industry, ivory-works, leather industry, pearl industry, etc.

Despite international ban on trade in products from endangered species, smuggling of fur, hide, horns, tusks, live specimen, etc. worth millions of dollars are being sold every year. Developing countries in Asia, Africa and Latin America are the richest biodiversity centers and wildlife products are smuggled and marketed in large quantities to some rich Western countries and also to China and Hong Kong where export of cat skins and snake skins fetches a booming business.

(iii) *Social Value:* These are the values associated with the social life, customs, religion and psycho-spiritual aspects of the people. Many of the plants are considered holy and sacred in our country like Tulsi (holy basil), Peepal, Mango, Lotus, Bael, etc. The leaves, fruits or flowers of these plants are used in worship or the plant itself is worshipped. The tribal people are very closely linked with the wildlife in the forests. Their social life, songs, dances and customs are closely woven around the wildlife. Many animals like cow, snake, bull, peacock, owl, etc. also have significant place in our psycho-spiritual arena and thus hold special social importance. Thus biodiversity has distinct social value, attached with different societies.

(iv) *Ethical Value:* It is also sometimes known as existence value. It involves ethical issues like "all life must be preserved". It is based on the concept of "Live and Let Live". If we want our human race to survive, then we must protect all biodiversity, because biodiversity is valuable.

The ethical value means that we may or may not use a species, but knowing the very fact that this species exists in nature gives us pleasure. We all feel sorry when we learn that "passenger pigeon" or "dodo" is no more on this earth. We are not deriving anything direct from kangaroo, zebra or giraffe, but we all strongly feel that these species should exist in nature. This means, there is an ethical value or existence value attached to each species.

(v) *Aesthetic Value:* Great aesthetic value is attached to biodiversity. No one of us would like to visit vast stretches of barren lands with no signs of visible life. People from far and wide spend a lot of time and money to visit wilderness areas where they can enjoy the aesthetic value of biodiversity and this type of tourism is now known as eco-tourism. The "Willingness to pay" concept on such eco-tourism gives us even a monetary estimate for aesthetic value of biodiversity. Eco-tourism is estimated to generate about 12 billion dollars of revenue annually, that roughly gives the aesthetic value of biodiversity.

(vi) *Option Values:* These values include the potentials of biodiversity that are presently unknown and need to be explored. There is a possibility that we may have some potential cure for AIDS or cancer existing within the depths of a marine ecosystem, or a tropical rainforest.

Thus option value is the value of knowing that there are biological resources existing on this biosphere that may one day prove to be an effective option for something important in the future. Thus, the option value of biodiversity suggests that any species may prove to be a miracle species someday. The biodiversity is like precious gifts of nature presented to us. We should not commit the folly of losing these gifts even before unwrapping them.

The option value also includes the values, in terms of the option to visit areas where a variety of flora and fauna, or specifically some endemic, rare or endangered species exist.

(vii) *Ecosystem Service Value:* Recently, a non-consumptive use value related to self-maintenance of the ecosystem and various important ecosystem services has been recognized. It refers to the services provided by ecosystems like prevention of soil erosion, prevention of floods, maintenance of soil fertility, cycling of nutrients, fixation of nitrogen, cycling of water, their role as carbon sinks, pollutant absorption and reduction of the threat of global warming, etc.

Different categories of biodiversity value clearly indicate that ecosystem, species and genetic diversity all have enormous potential and a decline in biodiversity will lead to huge economic, ecological and socio-cultural losses.

Biodiversity Value of Some Selected Organisms in Monetary Terms: A male lion living upto an age of 7 years can generate upto $ 515,000 due to its aesthetic value as paid by tourists, whereas if killed for the lion skin a market price upto $ 1,000 can be fetched.

In its lifetime a Kenyan elephant can earn worth $ 1 million as tourist revenue. The mountain gorillas in Rwanda are fetching $ 4 million annually through eco-tourism. Whale watching on Hervey Bay on Queensland's coast earns $12 million annually.

Tourism to Great Barrier Reef in Australia earns $ 2 billion each year. A typical tree provides $ 196,2150 worth of ecological services as oxygen, clean air, fertile soil, erosion control, water recycling, wildlife habitat, toxic gas moderation, etc. Whereas its worth is only about $ 590 if sold in the market as timber.

Biodiversity at International Level: Following the 1992 "Earth Summit" at Rio de Janeiro, it became evident that there is a growing need to know and scientifically name, the huge number of species which are still unknown on this earth. Roughly 1.5 million species are known till date which is perhaps 15% or may be just 2% of the actual number. Tropical deforestation alone is reducing the biodiversity by half a per cent every year. Mapping the biodiversity has therefore, been rightly recognized as an emergency task in order to plan its conservation and practical utilization in a judicious manner.

Terrestrial biodiversity of the earth is best described as biomes, which are the largest ecological units present in different geographic areas and are named after the dominant vegetation *e.g.* the tropical rainforests, tall grass prairies, savannas, desert, tundra, etc. The tropical rainforests are inhabited by teeming millions of species of plants, birds, amphibians, insects as well as mammals.

They are the earth's largest storehouse of biodiversity. Many of these species have developed over the time in highly specialized niches and that makes them more vulnerable to extinction when their natural home or niche is destroyed. About 50 to 80% of global biodiversity lies in these rainforests.

More than one-fourth of the world's prescription drugs are extracted from plants growing in tropical forests. Out of the 3000 plants identified by National Cancer Research Institute as sources of cancer-fighting chemicals, 70% come from tropical rain forests.

Very recently, extract from one of the creeping vines in the rainforests at Cameroon has proved effective in the inhibition of replication of AIDS virus. It is interesting to note that the common neem tree, so popular in tropical India, known for its medicinal properties has now come into lime light even in the Western temperate countries.

There is an estimated 1,25,000 flowering plant species in tropical forests. However, till now we know only 1-3% of these species. Needless to say, we must try in every way to protect our tropical rainforests.

The Silent Valley in Kerala is the only place in India where tropical rain forests occur. You may recall the case of Silent Valley Hydroelectric Project, which was abandoned mainly because it had put to risk our only tropical rain forest biodiversity.

Temperate forests have much less biodiversity, but there is much better documentation of the species. Globally, we have roughly 1,70,000 flowering plants, 30,000 vertebrates and about 2,50,000 other groups of species that have been described. There is a stupendous task of describing the remaining species which may range anywhere from 8 million to 100 million.

The below-given table shows the estimated number of some known living species in different taxonomic groups:

Living Species Estimates (World Resource Institute, 1999)

Taxonomic group	*Number*
Bacteria & Cyanobacteria	5,000
Protozoans (Single-celled animals)	31,000
Algae	27,000
Fungi (Molds, Mushrooms)	45,000
Higher Plants	2,50,000
Sponges	5,000
Jelly fish, Corals, etc.	10,000
Flatworms, roundworms, earthworms	36,000
Snails, Clams, Slugs, etc.	70,000
Insects	7,50,000
Mites, Ticks, Croaks, shrimps	1,20,000

Contd....

Taxonomic group	*Number*
Fish and Sharks	22,000
Amphibians	4,000
Reptiles	5,000
Birds	9,000
Mammals	4,000
Total	1,400,000

It is interesting to know that marine diversity is even much higher than terrestrial biodiversity and ironically, they are still less known and described. Estuaries, coastal waters and oceans are biologically diverse and the diversity is just dazzling. Sea is the cradle of every known animal phylum. Out of the 35 existing phyla of multicellular animals, 34 are marine and 16 of these are exclusively marine.

National Level Bio-diversity (Indian Biodiversity): Every country is characterized by its own biodiversity depending mainly on its climate. India has a rich biological diversity of flora and fauna. Overall six per cent of the global species are found in India. It is estimated that India ranks 10th among the plant rich countries of the world, 11th in terms of number of endemic species of higher vertebrates and 6th among the centers of diversity and origin of agricultural crops.

The total number of living species identified in our country is 150,000. Out of a total of 25 biodiversity hot-spots in the world, India possesses two, one in the north-east region and one in the western ghats. India is also one of the 12 mega-biodiversity countries in the world, which will be discussed later.

Biodiversity at Local or Regional Levels: Biodiversity at regional level is better understood by categorizing species richness into four types, based upon their spatial distribution as discussed below.

(i) Point richness refers to the number of species that can be found at a single point in a given space.

(ii) Alpha (α-) richness refers to the number of species found in a small homogeneous area.

(iii) Beta (β-) richness refers to the rate of change in species composition across different habitats.

(iv) Gamma (γ-) richness refers to the rate of change across large landscape gradients.

α-richness is strongly correlated with physical environmental variables. For example, there are 100 species of tunicates in arctic waters, 400 species in temperate waters and 600 in tropical seas. Thus, temperature seems to be the most important factor affecting α-richness of tunicates. β-richness means that the cumulative number of species increases as more heterogeneous habitats are taken into consideration. For example, the ant species found in local regions of north pole is merely 10. As we keep on moving towards the equator and thus add more and more habitats, the number of species of ants reaches as high as 2000 on the equatorial region.

India's Status as a Mega Diversity Nation: India is one of the 12 mega-diversity countries in the world. The Ministry of Environment and Forests, Govt. of India (2000) records 47,000 species of plants and 81,000 species of animals which is about 7% and 6.5% respectively of global flora and fauna. Distribution of species in some major groups of flora and fauna in India:

Group-wise Species Distribution

Plants	*Number*	*Animals*	*Number*
Bacteria	850	Lower groups	9979
Fungi	23,000	Mollusca	5042
Algae	2500	Arthropoda	57,525
Bryophytes	2564	Pisces (Fishes)	2546
		Amphibia	

Contd...

Plants	*Number*	*Animals*	*Number*
Pteridophytes	1022	Reptiles	428
Gymnosperms	64	Birds	1228
Angiosperms	15,000		204
		Mammals	372

Endemism: Species which are restricted only to a particular area are known as endemic. India shows a good number of endemic species. About 62% of amphibians and 50% of lizards are endemic to India. Western ghats are the site of maximum endemism.

Centre of Origin: A large number of species are known to have originated in India. Nearly 5000 species of flowering plants had their origin in India. From agro-diversity point of view also our country is quite rich. India has been the centre of origin of 166 species of crop plants and 320 species of wild relatives of cultivated crops, thereby providing a broad spectrum of diversity of traits for our crop plants.

Marine Diversity: Along 7500 km long coastline of our country in the mangroves, estuaries, coral reefs, back waters, etc. there exists a rich biodiversity. More than 340 species of corals of the world are found here. The marine diversity is rich in mollusks, crustaceans (crabs, etc.), polychaetes and corals. Several species of mangrove plants and sea-grasses (Marine algae) are also found in our country. A large proportion of the Indian Biodiversity is still unexplored.

There are about 93 major wet lands, coral reefs and mangroves which need to be studied in detail. Indian forests cover 64.01 million hectares having a rich biodiversity of plants in the Trans-Himalayan, north-west, west, central and eastern Himalayan forests, western ghats, coasts, deserts, Gangetic plains, Deccan plateau and the Andaman, Nicobar and Lakshadweep islands. Due to very diverse climatic conditions there is a complete rainbow spectrum of biodiversity in our country.

Biodiversity—Main Places: *Areas which exhibit high species richness as well as high species endemism are termed as hot spots of biodiversity.* The term was introduced by Myers (1988). There are 25 such hot spots of biodiversity on a global level out of which two are present in India, namely the Eastern Himalayas and Western Ghats. These hotspots covering less than 2% of the world's land area are found to have about 50% of the terrestrial biodiversity. According to Myers *et al.* (2000) an area is designated as a hotspot when it contains at least 0.5% of the plant species as endemics.About 40% of terrestrial plants and 25% of vertebrate species are endemic and found in these hotspots. After the tropical rain forests, the second highest number of endemic plant species are found in the Mediterranean (Mittermeier).

Broadly, these hot spots are in Western Amazon, Madagascar, North and East Borneo, North Eastern Australia, West Africa and Brazilian Atlantic forests. These are the areas of high diversity, endemism and are also threatened by human activities.

More than 1 billion people (about 1/6th of the world's population), most of whom are desperately poor people, live in these areas.

Any measures of protecting these hotspots need to be planned keeping in view the human settlements and tribal issues. Earlier 12 hot spots were identified on a global level. Later Myers *et al* (2000) recognized 25 hot spots.

Two of these hotspots lie in India extending into neighbouring countries namely, Indo-Burma region (covering Eastern Himalayas) and Western Ghats- Sri Lanka region. The Indian hotspots are not only rich in floral wealth and endemic species of plants but also reptiles, amphibians, swallow tailed butterflies and some mammals.

Global Hotspots of Biodiversity

Hotspots	Plant Species	Endemic Plants	%of Global Plants	Vertebrate Species	Endemic Vertebrates	%of Global Vertebrates
1. Tropical Andes	45000	20000	6.7	3389	1567	5.7
2. Mesoamerican forests	24000	5000	1.7	2859	1159	4.2
3. Caribbean	12000	7000	2.3	1518	779	2.9
4. Brazil's Atlantic Forest	20000	8000	2.7	1361	567	2.1
5. Choc/Darien of Panama Western Ecuador	9000	2250	0.8	1625	418	1.5
6. Brazil's Cerrado	10000	4400	1.5	1268	117	0.4
7. Central Chile	3429	1605	0.5	335	61	0.2
8. California Flonstic Province	4426	2125	0.7	584	71	0.3
9. Madagascar	12000	9704	3.2	987	771	2.8
10. Eastern Arc and Coastal Forest of Tanzania/Kenya	4000	1500	0.5	1019	121	0.4
11. Western African Forests	9000	2250	0.8	1320	270	1.0
12. Cape Floristic Province	8200	5682	1.9	562	53	0.2

Contd...

Hotspots	*Plant Species*	*Endemic Plants*	*%of Global Plants*	*Vertebrate Species*	*Endemic Vertebrates*	*%of Global Vertebrates*
13. Succulent Karoo	4849	1940	0.6	472	45	0.2
14. Mediterranean Basin	25000	13000	4.3	770	235	0.9
15. Caucasus	6300	1600	0.5	632	59	0.2
16. Sundaland	25000	15000	5.0	1800	701	2.6
17. Wallacea	10000	1500	0.5	1142	529	1.9
18. Philippines	7620	5832	1.9	1093	518	1.9
19. Indo-Burma						
Eastern Himalayas	13500	7000	2.3	2185	528	1.9
20. South-Central China	12000	3500	1.2	1141	178	0.7
21. Western-Ghats Sri Lanka	4780	2180	0.7	1073	355	13
22. South-western Australia	5469	4331	1.4	456	100	0.4
23. New Caledonia	3332	2551	0.9	190	84	0.3
24. New Zealand	2300	1865	0.6	217	136	0.5
25. Polynesia/Micronesia	6557	3334	1.1	342	223	0.8
Total	—	133,149	44.4	—	9645	35.3

Source: Myers *et al*, 2000.

(a) *Eastern Himalayas:* They display an ultra-varied topography that fosters species diversity and endemism. There are numerous deep and semi-isolated valleys in Sikkim which are extremely rich in endemic plant species. In an area of 7298 Km^2 of Sikkim about 4250 plant species are found of which 60% are endemic.

The forest cover of Eastern Himalayas has dwindled to about 1/3rd of its original cover. Certain species like *Sapria himalayana,* a parasitic angiosperm was sighted only twice in this region in the last 70 years.

Recent studies have shown that North East India along with its contiguous regions of Burma and Chinese provinces of Yunnan and Schezwan is an active centre of organic evolution and is considered to be the cradle of flowering plants. Out of the world's recorded flora 30% are endemic to India of which 35,000 are in the Himalayas.

(b) *Western Ghats:* It extends along a 17,000 Km^2 strip of forests in Maharashtra, Karnataka, Tamil Nadu and Kerala and has 40% of the total endemic plant species. Sixty-two per cent amphibians and 50% lizards are endemic to Western Ghats.

Forest tracts upto 500 m elevation covering 20% of the forest expanse are evergreen while those in 500-1500 m range are semi-evergreen. The major centers of diversity are *Agastyamalai Hills* and *Silent* Valley—the New Amambalam Reserve Basin. It is reported that only 6.8% of the original forests are existing today while the rest has been deforested or degraded, which raises a serious cause of alarm, because it means we have already lost a huge proportion of the biodiversity.

Although the hotspots are characterized by endemism, interestingly, a few species are common to both the hotspots in India. Some common plants include *Ternstroemia japonica,*

Rhododendron and *Hypericum,* while the common fauna includes laughing thrush, fairy blue bird, lizard hawk, etc. indicating their common origin long back in the geological times.

Challenges to Biodiversity: Extinction or elimination of a species is a natural process of evolution. In the geologic period the earth has experienced mass extinctions. During evolution, species have died out and have been replaced by others. However, the rate of loss of species in geologic past has been a slow process, keeping in view the vast span of time going back to 444 million years.

The process of extinction has become particularly fast in the recent years of human civilization. In this century, the human impact has been so severe that thousands of species and varieties are becoming extinct annually. *One of the estimates by the noted ecologist, E.O. Wilson puts the figure of extinction at 10,000 species per year or 27 per day!* This startling figure raises an alarm regarding the serious threat to biodiversity. Over the last 150 years the rate of extinction has escalated more dramatically. If the present trend continues we would lose 1/3rd to 2/3rd of our current biodiversity by the middle of twenty-first century. Let us consider some of the major causes and issues related to threats to biodiversity.

Habitats in Extinction

Destruction and loss of natural habitat is the single largest cause of biodiversity loss. Billions of hectares of forests and grasslands have been cleared over the past 10,000 years for conversion into agriculture lands, pastures, settlement areas or development projects. These natural forests and grasslands were the natural homes of thousands of species which perished due to loss of their natural habitat. Severe damage has been caused to wetlands thinking them to be useless ecosystems. The unique rich biodiversity of the wetlands, estuaries and mangroves are under the most serious threat today.

The wetlands are destroyed due to draining, filling and pollution thereby causing huge biodiversity loss. Sometimes the loss of habitat is in installments so that the habitat is divided into small and scattered patches, a phenomenon known as habitat fragmentation. There are many wildlife species such as bears and large cats that require large territories to subsist. They get badly threatened as they breed only in the interiors of the forests. Due to habitat fragmentation many song birds are vanishing.

There has been a rapid disappearance of tropical forests in our country also, at a rate of about 0.6% per year. With the current rate of loss of forest habitat, it is estimated that 20-25% of the global flora would be lost within a few years. Marine biodiversity is also under serious threat due to large-scale destruction of the fragile breeding and feeding grounds of our oceanic fish and other species, as a result of human intervention.

Poaching: Illegal trade of wildlife products by killing prohibited endangered animals *i.e.* poaching is another threat to wildlife. Despite international ban on trade in products from endangered species, smuggling of wildlife items like furs, hides, horns, tusks, live specimens and herbal products worth millions of dollars per year continues. The developing nations in Asia, Latin America and Africa are the richest source of biodiversity and have enormous wealth of wildlife. The rich countries in Europe and North America and some affluent countries in Asia like Japan, Taiwan and Hong Kong are the major importers of the wildlife products or wildlife itself.

The trading of such wildlife products is highly profit making for the poachers who just hunt these prohibited wildlife and smuggle it to other countries mediated through a mafia. The cost of elephant tusks can go upto $ 100 per kg; the leopard fur coat is sold at $ 100,000 in Japan while bird catchers can fetch upto $ 10,000 for a rare hyacinth macaw, a beautiful coloured bird, from Brazil. The worse part of the story is that for every live animal that actually gets into the market, about 50 additional animals are caught and killed.

If you are fond of rare plants, fish or birds, please make sure that you are not going for the endangered species or the wild-caught species. Doing so will help in checking further decline of these species. Also do not purchase furcoat, purse or bag, or items made of crocodile skin or python skin. You will certainly help in preserving biodiversity by doing so.

4

Elements at Work

Role of Water

Water covers about 73% of the earth's surface and provides the most extensive medium for aquatic life. It constitutes about 73-90% of protoplasm and is most essential for almost all the metabolic and chemical reactions. Water also acts as a major controlling factor of the organisms.

Properties of Water: Water has several unique qualities from ecological point of view which render it as the most suitable medium for the living beings. HENDERSON (1924) in his book—"The fitness of the Environment" has discussed these unusual qualities of water. Some of these properties are discussed under the following headings:

Water as a Solvent: Water is an universal solvent. No other liquid is comparable to water in this respect. Because of its solvent property, water contains several minerals and gases needed for the building up of protoplasm.

The exchange of oxygen required by the living organisms and the carbon dioxide produced takes place in the watery medium. In fact water forms the only appropriate solvent for solutes required by the living beings.

Thermal Properties: Water has high specific and latent heat. These together combine to minimize the range of temperature changes:

(i) *Specific Heat:* It is the number of calories required to increase the temperature of one ml. water through 1°C. The specific heat of water is high so that it has the ability to withhold amount of heat with slight variation in temperature. A large amount of heat can be received or given out from its surface to the surrounding atmosphere without any appreciable change in the temperature. Thus the high specific heat of water help in maintaining the thermal stability of the water body (pond, lake or seas).

(ii) *Latent Heat:* Latent heat is the quantity of heat required to change a substance from the solid to the liquid or from liquid to the gaseous state without any change in the temperature. Water has latent heat. This property is of great ecological significance. Water has a relatively high freezing point (0°C). Because of this, large amount of heat is given up before water turns to ice and further due to restricted stirring oceans and lakes freeze only at the surface.

The temperature of the medium, therefore, can drop only to 0°C in freshwater ponds and lakes and to a few degrees lower in the ocean (upto –2.5°C). The biological reactions of a large number of plants and animals can still go on perfectly well at temperatures down to freezing point of water.

Role of Land

In case of water, surface tension may be defined as the lateral attraction amongst the molecules of water by which they are held together. Water has the highest surface tension of common substances except mercury. The high surface tension of water results in the formation of vacuoles within the cells and outside the cells and movement of water into and through the organisms as well as the rise of ground water into the soil.

Viscosity: The viscosity of water is high enough to protect the aquatic animals from mechanical disturbances. Because of high viscosity of water planktonic organisms float passively on water surface without any elaborate swimming organs.

Density: Water is heaviest at 4°C. The freezing point of freshwater is 0°C and that of sea water is about –2.5°C so that ice can float on its surface. Thus deep lakes and ponds do not freeze upto the bottom where the animals live comfortably.

Heat Conduction: Water has the highest heat conductivity among the liquids. This heat conducting property of water helps in the maintenance of a constant temperature. The heat generated by the activity of a cell is conducted immediately by water to other cells and is evenly distributed over the entire body.

Salinity: The salinity of marine water is rather constant in all oceans being about 3.5%. This poses least restriction on the movement of marine animals from one place to another in the ocean. On the other hand, the salinity of freshwater varies greatly and sometimes acts as a limiting factor in the distribution of animals. Some salt lakes have a salinity as high as 25-30%. This greatly restricts life in them.

Transparency: Transparency of water is affected by the suspended impurities which also act as a limiting factor. Penetration of light also depends on the angle of incidence and the nature of the wave lengths. Light with longer wave lengths is absorbed by the upper layers of water whereas light with short wavelengths penetrates deep into water. Ultraviolet rays may reach upto a depths of 1600 meters. Below the light zone the ocean is completely dark except some light emitted by the luminiscent organisms.

Pressure: Pressure increases with increased depth of water at the rate of one atmosphere for every 10 meters of descent. Animals inhabiting the floor of deep-sea areas at depth of 10,500 meters are exposed to pressure of about one ton per square centimeter. The aquatic animals at such a depth will experience a change in shape and size of air filled chambers within their body. The swim bladder of a fish will decrease to one-half of its surface volume at a depth of 10 meters.

Therefore, fishes with swim bladders are limited with respect to vertical distribution. The mammals have to come to the surface periodically to expel the accumulated CO_2 and replenish their supply of oxygen. Most animals living at great depths lack air-filled cavities. The cavities or spaces that may exist are filled with fluids.

Hard and Soft Waters: Water is hard when it does not give foam with soap. Hard water contains soluble salts of calcium and magnesium such as chlorides, sulphates and bicarbonates. When these salts are absent, the water is termed as soft. The hardness of water is of two types—temporary hardness and permanent hardness. Temporary hardness is due to the presence of bicarbonates of calcium and magnesium. On boiling temporary hardness is removed and water becomes soft due to the conversion of soluble bicarbonates into insoluble carbonates. Permanent hardness is due to the chloride and sulphates of calcium and magnesium. This can be removed only by certain chemicals.

Composition of Natural Waters: The amount of dissolved materials in natural waters is usually different from the fluids of the organisms. The differences resulting in the osmotic pressure of the organisms and their environment raises a problem in regard to water exchange. In order to determine whether the water will tend to enter or to leave the organism, it is essential to know the composition of the surrounding water.

From the table it is clear that hard freshwater contains more dissolved salts than soft freshwater. In sea-water chloride and sodium are the most abundant ions whereas in hard freshwater carbonate and calcimines are most abundant. In soft freshwater, calcium and carbonate ions are relatively less concentrated than sodium and chloride.

Showing Composition of Some Natural Waters (After Baldwin, 1948)

Water	*Na*	*K*	*Ca*	*Mg*	*Cl*	*SO_3*	*CO_4*	*Total gm/litre*
Soft freshwater	0.016	–	0.010	–	0.019	0.107	0.012	0.065
Hard water	0.021	0.016	0.065	0.014	0.041	0.125	0.119	0.30
Sea water	10.7	0.38	0.42	1.31	19.3	2.69	0.073	34.9

Water Problem in Different Habitats: The concentration of water divides the environment into aquatic and terrestrial environments. Organisms living in these environments face different problems in maintaining proper water balance and have evolved various contrivances.

Water Problem in Aquatic Environment: Depending upon the concentration of salts, the aquatic medium can be divided into marine, estuarine, and freshwater habitats. Animals living in any of the above mentioned habitats have to face the problem of water balance inside their body tissues, except when the osmotic pressure inside the tissues of the organisms is equal to that of surrounding medium.

The freshwater animals suffer from endosmosis, where the water tends to enter the body through all the possible surfaces which are exposed to the surrounding medium and tends to dilute the tissue fluid. Various methods are exercised by different animals to get rid of this excess water. Protozoans possess contractile vacuoles which collect excess of water from body fluid and burst out of the body releasing water to the exterior. The freshwater fishes have an exoskeleton of chitinous scales which prevents absorption of water through general body surface. These excrete hypotonic urine with excess of water. On the contrary, most of the marine vertebrates have their body fluids even more watery than sea-water. Therefore, their survival in sea-water is possible only if they are able to imbibe water and excrete salts.

Aquatic organisms are commonly adapted to live within a specific range of salt concentration. Usually the limitations of these adaptive mechanisms restrict the animals either to fresh or salt-water. These could tolerate only narrow fluctuations of salt concentration. Such animals are known as stenohaline. But animals are found to tolerate wider ranges of salt concentration. These are known as euryhaline. Animals of estuarine water (tidal water) are euryhaline and can withstand greater fluctuations in salt concentration. Certain animals are stenohaline as well as euryhaline during different periods of their life span (*e.g.* salmon and eels). Salmons hatch in the warm-waters of certain rivers, where salt concentration is very low. As juveniles they migrate to the ocean. At maturity, they again enter the rivers of their origin.

Water Problem in Terrestrial Animals: Terrestrial animals which face water problem are forced to conserve water at any cost. During the course of evolution these have developed various adaptations in this direction. Some of these are summarised as under:

1. *Burrowing Habits:* Animals like earthworms, frogs and toads burrow in the ground and reach the moist earth where they find optimum living conditions. Lung fishes that live in ponds burry themselves when water dries up.
2. *Nocturnal Habit:* Life in day time necessitates the animals to face hot sun, high temperature, low humidity and a high rate of evaporation. In order to avoid this, many animals change over to nocturnal habits.

3. *Dry Impervious Skin:* The main source of evaporation of body fluids for all animals is their skin. The animals have developed a hard, dry impervious skin with scales, spines, scutes, shields as specialised outer covering which almost seal their body fluids from evaporation.
4. *Sweat Glands:* There is either a reduction or absence of sweat glands in mammals that face dry conditions.
5. *Respiratory System:* The terrestrial animals have discarded the aquatic gills once they moved into land. This has been substituted by internal lungs in most of the animals and tracheal system in insects. The loss of water is very less in tracheal system and seems to be the most ideal mechanism for conservation of water.
6. *Suspended Animation:* Diapause or the retardation of growth and development during dry season is common among many insects. A large number of animals aestivate. Tradigrades and rotifers withstand long drying and can come back to life once conditions improve.
7. *Migration:* Birds and mammals migrate to far off places during the dry seasons and return to their original abode when conditions become favourable.
8. *Excretion:* The removal of wastes involves much loss of water. However, animals save water by changing their excretory products to much concentrated solid uric acid as in reptiles, insects etc. Insects, birds and reptiles further deposit dry faces and thereby reduce wastage of water.
9. *Hygroscopic Arrangements:* Some animals can even absorb moisture through the skin, *e.g.* Grasshopper (Ludwig).
10. *Water from Food:* Many herbivores depend upon succulent vegetation for their water needs while

carnivores entirely depend on the body fluids and blood of animals. Insects like (Cryptotermes) are able to live even on the moisture obtained from dry wood.

11. *Water from Metabolism:* Animals especially insects and desert rat utilise the water of metabolism. This water is obtained by the break-down of carbohydrates and the oxidation of hydrogen. Lipids are a good source of metabolic water as they contain large amount of hydrogen.

Water Problem in Amphibious Situation: Many aquatic organisms have made attempts to invade and colonise land. But only a few groups of animals have succeeded in colonising the land. This success is due to their ability to secure and retain sufficient water during adverse conditions. However, some plants and animals got success in coming out of water for short periods. Such animals have developed certain methods of using the advantages of both the air and the watery medium. The best known examples are amphibians. Though they have not completely adapted themselves for terrestrial environment, they are relatively independent of water to some extent. They have to re-enter water or visit damp places at intervals to keep their skin moist. Most of them need water for reproduction.

Organisms living in swamps and pools are able to tide over the periods of drought by various means. Certain species of fish in India can live in wet grass for about 60 his. Some animals burrow in mud and remain dormant during the dry period. The African lungfish, Protopterus, constructs a special mud cocoon and remain inside the cocoon during the dry spell.

Daphnia (a crustacean) reproduces by parthenogenesis in a temporary pool. The young ones on maturation also reproduce the same way. Later, when the pool tends to dry up, the males make their appearance. Now females produce a different type of egg that requires fertilization. Only one such egg is formed by each female. This egg remains as a resistant egg and survives prolonged periods of drought. Certain protozoans, sponges, bryozoans also overcome adverse conditions by producing spores persistent eggs or cysts.

Tidal zones are also known as amphibious. The fauna and flora of tidal zone consists of sessile forms such as the common sea-weeds, barnacles, mussels and clams. Attached forms with shells like barnacles, mussels and oysters close up their openings during the period of low tide and many of them can live in this condition for a very long period without injury.

Effect of Humidity on Growth and Distribution of Animals : An important water relation is the relative humidity of air. It is the percentage of the amount of water vapours that air could hold at saturation at the existing temperature.

Relative humidity determines the occurrence of a particular kind of vegetation and so also the distribution and occurrence of animals Amphibians, insects and terrestrial invertebrates are greatly affected by the presence or absence of moisture.

Silverfish (Lepisma saccharina) requires 80-90% humidity for reproduction. The newly hatched nymph die at a humidity of less than 85-90%. In flour beetle, Tribolium, the larval stage is accelerated by an increase in relative humidity, but the duration of the egg and pupal stages are unaffected by changes in humidity.

When moisture conditions of the environment become adverse, the animals concerned either die or migrate to longer or shorter distances. Droughts greatly reduce the occurrence of some species temporarily or for long periods by decimating the population. Locomotor activity, as well as feeding and other reactions are controlled by humidity in certain animals.

The Precipitation: The moisture falling on an area in liquid, vapour of frozen form is designated as precipitation. Precipitation depends upon season, wind, air, pressure and temperature. Rainfall is the most common form of precipitation. The amount and manner of rainfall is of considerable ecological importance. Light drizzle is of little importance as very little moisture penetrates the soil because much of it evaporates rapidly. On the contrary, gentle steady rain is most effective because much of it penetrates the soil. Torrential rains are most disastrous because sustained torrential rainfall leads to extensive soil erosion and flooding.

Rainfall is not uniformly distributed. Mean annual rainfall is influenced by topography and mass air movements. Himalayas in the north interfere with the monsoons coming off the Bay of Bengal. This causes excessive rainfall on the northern slopes of the Himalayas.

Seasonal distribution of rainfall is more important to organisms than average annual rainfall. Considerable difference exists in a region with 50 inches of rainfall evenly distributed throughout the year and a region in which almost all the 50 inches of rain falls during a few months period. This is characteristic of tropical and subtropical climates where the organisms face along period of drought. An alternation of wet and dry seasons influences the reproductive activity and cycles of organisms.

Hail, sleet and snow are forms of frozen rain. Hail is in the form of balls of frozen moisture. These may be of considerable size and weight. Hail may cause considerable damage to agriculture. Sleet and snow form destructive glaze on trees, shrubs and grass.

The weight of that is great enough to snap branches and cause considerable damage. Since animals depend on the vegetation for food and shelter, and the vegetation is directly dependent upon the quantity and distribution of precipitation in terrestrial areas, all the living components of the environment depend on precipitation either directly or indirectly.

Significance of Soil

ALLEE, PARK and PARK have defined soil as the shallow upper layers of the earth that are formed by the weathering of underlying rocks in association with organise matter and with living organisms and has become a suitable habitat for plants and animals. KNIGHT (1956) defined soil as the mineral material that exists in solid or unbroken form such as boulders and gravels or finely divided particles of mineral matter such as sand, silt or clay depending upon the texture. TRESHOW (1970) gave a more appropriate definition of soil. He defined soil as a complex physical-biological system providing support, water, nutrient and oxygen for the plants. Soil is an important habitat for a large number of terrestrial organisms, either burrowing or living on its surface. Its organisms exhibit a complete food chain with herbivores, carnivores and the predators and prey. The study of soil is known as pedology.

Soil Formation: Soil originates from rocks. It develops gradually by the fragmentation and corrosion of rocks and with the accumulation of, organic matter. Soil formation may require 2,000 to 20,000 years but soil differentiation from the parent material may take place in a short time of 30-50 years.

Soil Forming Rocks: There are three main kinds of soil formind rocks-igneous, sedimentary and metamorphic rocks. The main characteristics of these rocks are given in the table.

Characteristics and Mineral Composition of the Principal Soil forming Rocks (After WALLWORK 1970)

	Class of rock	*Mode of formation of rock*	*Type*	*Description of rock types*	*Principal minarals of soil forming rocks*
I.	Igneous rocks	Formed by the cooling of molten megma	1. Granite	Coarse to medium grain, light in colour	Quartz, felspar, mica, iron oxides, amphibole, etc.
			2. Diorite	Coarse to medium grain, grey to dark in colour	Amphibole, iron oxide, biolite and feldspar

Contd...

Class of rock	*Mode of formation of rock*	*Type*	*Description of rock types*	*Principal minerals of soil forming rocks*
		3. Basalt	Dense to fine grain, dark to black in colour	Felspar, iron oxide, biolite, pyroxene, etc.
II. Sedimentary rocks	Formed by deposition of weathered minerals which are derived from igneous rocks	1. Shales	Light to dark in colour, thinly laminated	Clay minerals and quartz
		2. Sand stone	Light to red in colour, porous and granular	Quartz, claym-inerals, iron oxides, calcium carbonate, etc.
		3. Line-stone	Light grey, red-brown or black in colour, fine grained and compact	Calcite, dolomite, iron oxides, clay minerals
III. Metamorphic rocks	Formed by change of preexisting rocks, *i.e.* ig—neous or sedimentary through heat and pressure	1. Gneiss	With light and dark bands	Much as in granite from which it is formed
		2. Schist	Foliated	Consists of basalt or shale from which it is formed
		3. Slate	Grey to black in colour, compact and uniform texture	Similar to shale from which it is formed
		4. Quartzite	Light to brown in colour, compact and uniform texture	Simmilar to sand stone
		5. Marble	Light, red, green or black in colour, compact, fine to coarse texture	Calcite, dolomite formed from limestone

Process of Soil Formation: The process involved in the soil formation can be studied under following heads:

Weathering: It is the mechanical breakdown of rock material into smaller pieces. The mechanical breakdown of rock is brought about by strong winds, extreme temperature fluctuations, shearing effect of ice sheets, running water and root growth.

Chemical Weathering: It includes the chemical process of oxidation, reduction, hydration, hydrolysis and carbonation, etc. By the action of these processes soil materials go into the solution with rain water. This water percolates through the ground as soil water combines with CO_2 to form carbonic acid which combines with calcium and magnesium of the rocks to form their carbonates. Aluminium and silicon minerals are converted into clay and iron compounds are either reduced to ferrous state or are oxidised to ferric state. In this way rock is broken down into fine material which either remains at its place or may be transported to other places by wind and water.

Plants are also responsible for soil formation. The roots secrete certain acid which mixes with the ground water and causes the rocks to disintegrate rapidly.

Products of Weathering and Soil Types: The soil formed by weathering of soil-forming rocks is called embryonic or primary soil. It may mature into any of the following types of soils:

(i) *Residual Soil or Sedimentary Soil:* This is the mature soil that lies immediately over the parent rocks.

(ii) *Skeletal Soil or Secondary Soil:* This is partly weathered soil in which maturation has not taken place.

(iii) *Secondary Soil or Transported Soil:* In this type of soils, the weathered parent material has shifted to different places by the agency of glacier, streams and rivers, the gravitational forces as land slides, wind, sand storms, standing water and wave action and the oceanic wave.

The soils are also classified on the basis of profile development into zonal, intrazonal and azonal soils.

(i) *Zonal Soil:* In these soils climate is the primary factor in the process of soil formation. The fully developed soils of a climatic zone tend to be very similar.

(ii) *Intrazonal Soils:* Such soils are the result of local geological conditions that preclude the normal pattern of soil formation. These soils are of two types:

(a) *Hydromorphic Soils:* which are formed in areas in areas where the soil becomes water-logged during soil formation. Such soils are not aerated, and the normal aerobic breakdown of detritus cannot take place. These soils are rich in organic materials.

(b) *Halomorphic Soils:* are formed in arid and semiarid regions due to water logging. Such soils have high salt contents.

(iii) *Azonal Soils:* These are inherited soils, Alluvial soil is the best example of azonal soils. This type of soil is rich in colloids. Since it is not leached, it may be rich in nutrient materials, but lacks humus.

Mineralization: Organic matter of dead and decaying organisms also contributes to soil formation. The lichens and big trees die and their dead parts, the roots, fallen leaves and dried twigs get mixed with the rock material. To this are added the remains of animals and their excreta.

All organic plant debris fallen recently to the ground is called litter. It is composed of dead leaves, twigs, woods, dead roots and various plant roots. The litter is decomposed by combined efforts of soil microbes like bacteria, actinomycetes and other fungi.

The resultant products of decomposition include various kinds of inorganic and organic nutrients. These all get incorporated into the mineral particles of soil which now becomes dark in colour. This decomposed and black coloured organic matter which is added to the soil is called humus and the formation of humus is called humification. Gradually, all the humus is decomposed into simple compounds like CO_2, water, minerals and nutrients by the process of mineralization.

Sources of Chemical Nature of Litter: Forests, grasslands and aquatic plants are the main sources of litter. In deciduous forests huge amounts of litter are produced annually, whereas in evergreen forests the leaf fall is gradual. The chemical nature of litter depends upon the type of vegetation.

The tree litter of forests consists of calcium, potassium, magnesium, iron, manganese, silicon, copper, aluminium, phosphorus, nitrogen, cellulose, hemicellulose, lignin, fats, resins and waxes. Coniferous litter and evergreen types of litter have lower amounts of calcium than the deciduous litter. Indian tropical forests produce approximately 5 tons of litter per hectare per year.

Soil Profile: A cross-section of a well developed soil shows more or less distinct layers called horizons. The horizons may differ in thickness, colour, texture, structure, porosity, acidity and composition. This sequence of horizons from the surface downwards is called the soil profile. In a typical soil profile following horizons or layers can be recognised.

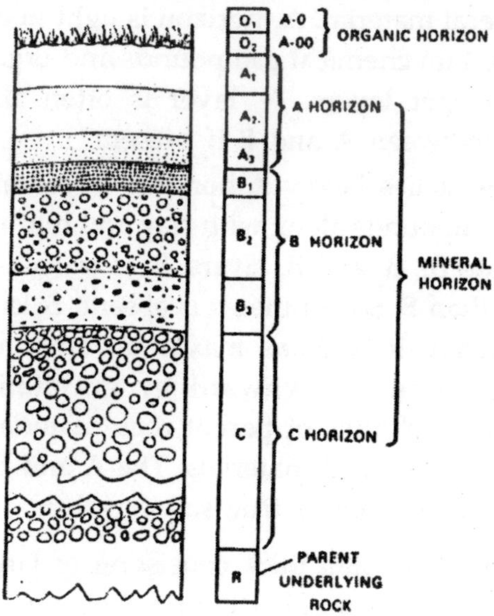

A generalized forest soil profile showing various zones.

A-horizon: This is the topmost soil in which the vegetation grows and is the scene of biological activities. It is darker in colour and rich in humus. Its thickness varies upto 10 ft. It can be further differentiated into the following layers.

A-oo: This is the top layer consisting of freshly fallen litter *i.e.* leaves and organic debris which is still intact or partially decomposed.

A-o: Below A-oo lies A-o layer. It contains organic matter in variable stages of decomposition. The upper portion of A-o layer contains organic matter in the initial stages of decomposition. This is also known as F layer or fermentation level. The lower portion of A-o horizon is also known as H or humus level. A-o horizon is very rich in life. A-o and A-oo layers of A-horizon are found under thick forests but may be absent under grasslands.

Below the upper layers of A-horizon is the first mineral layer in which three subdivisions, *viz.* A_1, A_2, A_3 may be present. A_1-horizon is dark in colour as organic debris is mixed here with the mineral material. A_2-horizon is light in colour due to leaching of certain chemical compounds and organic material down into deeper layers. A_3 layer is often absent and is intermediate between A and B if present.

B-horizon: It lies below A-horizon. It contains iron and aluminium compounds along with clay and humus. It can also be divided into B_1, B_2 and B_3 layers. B_1 is transitional to B but more like A, than B. Sometimes it is absent. B_2 is usually dark coloured horizon. It contains maximum amount of leached material transported downward by gravitational water. B_3 contains large chunks of parent rock material along with some quantity of leached materials. The A and B horizons of soil together are known as true soil or solum.

C-horizon: It is often thick consisting of large masses of weathered mineral material. The upper portion of C-horizon may become glued due to accumulation of water.

D-horizon: This is the lowest horizon of the soil profile. It comprises of unweathered parent rock material. In India we find the alluvial soils in the submountain regions along the base of Himalayas in W. Bengal, Bihar, U.P. and Punjab; in M.P. in the valleys of Narmada and Tapti; and Cauvery, Krishna and Godavari.

Laterites are found on the summit hills of Deccan, M.P., Eastern Ghats, parts of Orissa, Assam, Kerala and Mysore. The black cotton soils are found in Maharashtra and West Bengal. The red soils are found in large areas of North East Andhra and Mysore. Besides there are the mountain soils, soils of the deserts and arid regions, peaty and organic soils. Peaty soils are found in Kerala and are black and heavy and mostly acidic.

Soil Texture: Since soil influences the flora and fauna of the area, its texture is of considerable ecological interest. Texture is determined by the size of constituent particles which have been named differently by the International Society of Soil Science.

S.No.	*Name of Particles*	*Diameter of Particles*
1.	Gravel	2.00 mm. and more
2.	Coarse sand	2.00 mm. to 0.2 mm.
3.	Fine sand	0.20 mm. to 0.02 mm.
4.	Silt	0.02 mm to 0.002 mm.
5.	Clay	Below 0.002 mm.

Soils formed with various integrations of soil particles are of the following main types:

(i) Sandy soils mainly consist of sand particles. These are loose, dry and poor in nutrients. The water holding capacity of such soils is poor.

(ii) Clay Soils chiefly consist of clay particles. The clay particles are of colloidal dimensions. They have high plasticity and possess high water holding capacity. Clay particles have very small interspaces between them so that neither water nor air can circulate freely. Such soils on getting water become waterlogged. Thus they are not suitable for plant growth.

(iii) Loam soils have sand, silt and clay particles in more or less equal proportions. Such soils are the most suitable for plant growth.

(iv) Sand loams soils are those soils in which sand particles predominate..

(v) Clay loam soils have a predominance of clay particles. Both sandy and clay loam soils are suitable for plant growth.

(vi) Silt loam soils have predominance of silt. On getting water, silt loam becomes water-logged with poor air circulation. Such soils are not suitable for plant growth.

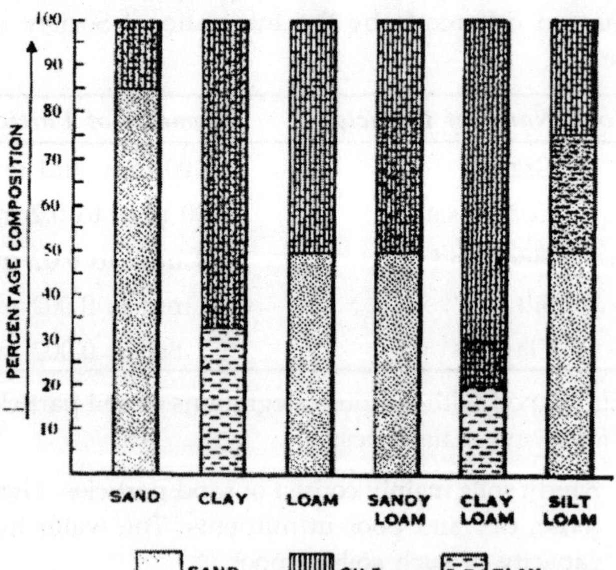

Diagrammatic representation of soil classification based on percentage of textural components of sand, silt and clay

Percentage of Sand, Silt and Clay in Main Soil Types

S.No.	Soil Class	Percentage		
		Sand	*Silt*	*clay*
1.	Sandy soil	85-100%	0-15%	0-15%
2.	Clay	0-59%	50-0'%	31% or more
3.	Loam	23-52%	28-50%	7-27%
4.	Sandy loam	50-80%	0-50%	0-20%
5.	Clay loam	20-80%	15-53%	20-30%
6.	Silt loam	0-50%	50-88%	0-27%
7.	Silt clay loam	0-20%	40-73%	27-40%

U.S. Deptt. of Agriculture—Forest Service Hand Book-Hand Book on Soils, Washington D.C., 1960. Most soils are mixtures of the above mineral particles. Based on the proportions of various particles contained in them, soils can be grouped into the following five groups:

1. *Coarse Textured Soils:* These are loose soils consisting mainly of sand and gravel. These retain only a small amount of moisture and supply some nutrients to the plants.
2. *Moderately Coarse Soils:* These include sandy loam to very fine sandy loam.
3. *Medium Textured Soils:* These are mixture of sand, silt and clay. They hold enough water and nutrients.
4. *Moderately Fine Textured Soils:* These are high in clay. These are moderately sticky and plastic when wet and form a crust on the surface if the organic matter in low.
5. *Fine Textured Soils:* These contain 40% clay. These may be sticky and plastic when wet and hold considerable water and nutrients but have limited internal drainage.

Soil Structure: The primary soil particles *i.e.* sand, silt and clay, usually occur in the form of aggregates. The arrangement of these individual particles into defined patterns is called soil structure. Natural aggregates are called peds whereas artificially formed soil mass is termed clod. Broken peds are called fragments. The soil structure is of the following four types:

1. *Prism-like:* These particles are prism-like but without rounded surface.
2. *Plate-like or Platy:* These are plate-like particles. Sometimes plates are overlapped.
3. *Block-like:* In this case the peds are cube-like with flat or rounded faces. When the faces are flat and the edges sharp and angular, the structure is termed angular blocky. When the faces and the edges are mainly rounded it is called sub-angular blocky.
4. *Sphere-like or Spheroidal:* These are rounded or spheroidal and smaller in size. The aggregates of this group are termed granular. These are less porous. When the granules are very porous, they are called as crumby.

Soil structure is influenced by texture, air, moisture, organic nutrients, micro-organisms, root growth and soil chemical nature. The best structure for favourable physical properties of soil are crumby and granular.

Major Soil Groups: There are four major soil groups in India, *viz.* Alluvial, Black (regur), Red Laterite and Lateritic soils.

Alluvial Soil: This covers about 142 million hectares of the land. The Gangetic plains, the Gujarat plain, the coastal tracts and deltas are the most extensive alluvial soils. They are the most productive of Indian soils, contributing the largest share of agricultural wealth of the country. Generally, it is deficient in N_2 and humus.

Black Soil: This forms the second largest group. It is developed mainly on the Deccan traps of Maharashtra, Madhya Pradesh and Kathiawar. They can be classified on the basis of depth. Deep black soil occurs in the Godawari and Tapti valleys, while shallow black soil is found in the eastern part of the Satpura plateau, extending southwards towards Nagpur.

Medium depth black soil called 'regur' is the most widespread. Because of its hydrology and climatic conditions of the environment, the medium and deep black soils are very suitable for cotton cultivation. Areas of regur are credited with high fertility and do not require manuring for long periods, but irrigation is essential.

Red Soils: These are developed mainly on the Archaean crystalline rocks of the Southern part of Deccan, and also on the sedimentary rocks of the north eastern mountains. Depth and fertility are found to vary considerably but the red soils are suitable for a variety of crops, if properly irrigated. They are generally red, often grading into brown, chocolate and yellow. The red colour is partly due to a high percentage of iron content. Found in Tamil Nadu, Karnataka, Andhra Pradesh, Santhal Pargana (Jharkhand) and some parts of West Bengal, Assam, U.P., and Rajasthan.

Lateritic Soils: These occupy about 4% of the total land area on the hills and plateaus of the north eastern states, Madhya Pradesh and the Deccan. Because of intensive leaching, typical lateritic soils are generally infertile and of little value for crop production. Also found in Eastern Ghats, Orissa, Assam, Karnataka and Malabar. Besides, there are some other locally important soils:

Forest Soils or Mountain: These cover an area of 285,000 sq.km. Such soils in the Darjeeling district consist of well developed horizons and are very suitable for the growth of tea plants. Characteristic deposition is organic matter derived from the forest.

Arid and Desert Soils: In western Rajasthan and adjoining dry areas desert soils cover about 142,000 sq.km. Excellent melons are now grown even on sand dunes.

Saline and Alkaline Soils: Such soils cover about 2.5 million hectares in U.P., Punjab and Harayana. Also found in Rajasthan.

Peaty Soils: Such soils remain submerged during the monsoon, and occur mainly in Kerala.

Skeletal Soils: So designated because of the thickness of the soil and lack of characteristic development in them. Such soils usually occur over the sand stones of the Vindhyachal formation. The colour is pale brown to dark brown. The poor nutrient supply from such soils sustain only a hardy type of natural vegetation and the soils have low value for cultivated crops. According to Charles E. Kellog, chief of the Soil Survey, U.S.A. Soil is the thin film between earth and the sky that supports all living and non-living things.

Soil Colour: Soil colour is of extreme importance for the identification of soil. In temperate regions, the dark coloured soils are rich in organic matter than the light ones. Well drained soils may range between very pale brown to dark brown and black, depending upon the organic matter content. Red and-yellow soils are due to the presence of iron oxides, and the bright colour of the soil indicates good drainage and good aeration. Well drained yellowish sands contain a small amount of organic matter and iron oxide. Red and yellow soils are due to the presence of iron oxides except the colour of the parent material. White and greyish soils are rich in quartz, carbonates of line and magnesium, gypsum and various compounds of ferrous iron. Reddish brown or red soils are less fertile than dark soils but are more fertile than white or yellow grey soils.

Physical Properties: Soil possesses many characteristic physical properties which can be studied under the following headings:

Soil Density: Though the average density of soil is 2.65 gm per ml., it varies greatly depending upon the degree of weathering.

Porosity: The spaces present between soil particles in a given volume are called pore-spaces. The percentage of soil volume occupied by the pore spaces is called porosity of the soil. The porosity of soil depends upon the texture, compactness and organic contents of the soil. The increase in organic matter increases porosity of the soil.

The pore-spaces are of two types *viz.* micro-pore spaces and macro-pore spaces. The macro-pore spaces (capillary pore spaces) can hold more water but restrict the free movement of water and air. The macro-pore spaces (non-capillary pore spaces) allow free movement of water and air in the soil as they have little water holding capacity.

Role of Soil

Soil permeability depends on the size of pores. It is higher for the loose soil which has large number of macro-pore spaces and is low in compact soil with micro-pore spaces.

Soil Temperature: Soil temperature is effected by its colour texture, water contents, slope, altitude of land and by climate and vegetation. Evaporation of water from soil makes it cooler. Black soils absorb more heat as compared to white soils. Sandy soils absorb more heat but radiate it out quickly at night than the loamy or clay soils.

Germination of seeds, normal growth of roots and activity of soil organisms require specific temperature.

Soil Air: Soil air is essential for the respiration of soil micro-organisms and underground parts of the plants. Compared to atmospheric air, soil air is slightly poorer in oxygen. This is due to the metabolic activities of roots and soil organisms.

The oxygen present in the soil air is an important factor in soil fertility since it is necessary for the breaking down of insoluble minerals into soluble salts. It is essential for soil organisms whose activity increases the soil fertility. In the absence of oxygen seed germination, root growth, water absorption by root hairs etc. are slowed down and ultimately altogether cease. Under deficient supply of oxygen, due to anaerobic respiration, production of toxins takes place. Well aerated soils are free from such toxic substances.

Soil Water: Soil retains large quantities of water. An ordinary garden soil contains about 25% water by volume. The principal source of soil water is rain. After a heavy rainfall, the water is retained by soil in various forms. On the basis of water retained by soil, it has been classified as gravitational water, capillary water, hygroscopic water and combined water.

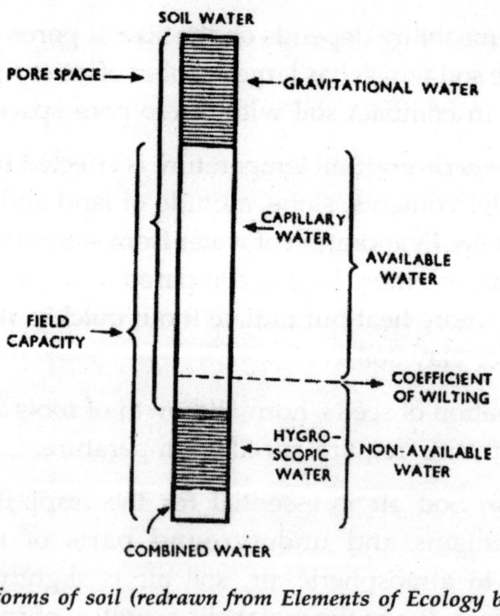

Different forms of soil (redrawn from Elements of Ecology by Clarke).

1. *Gravitational Water:* This is the free water which due to the action of gravity moves downwards into the soil until it reaches the water-table. This water is of little direct value to the plants.

2. *Capillary Water:* This water fills the spaces between non-colloidal soil particles and forms films around them. This water is held by the capillary forces around and between the particles and is of greatest importance to plant life.
3. *Hygroscopic Water:* The water that occurs in the form of thin film on the soil particles is termed as hygroscopic water. This water is not available to the plants.
4. *Combined Water:* A small portion of the soil water is chemically bound with soil materials. It is known as combined water.

Field Capacity: It is the total amount of capillary, hygroscopic and combined water plus the water vapour. It is the maximum amount of water that the soil can hold after the gravitational water has drained away.

Available and Non-available Water: The water that can be absorbed by the plants is termed available water. The water which the soil, does not yield to the plants is known as non-available water.

Wilting Point: When the percentage of moisture that is left in the soil at the time at which the plants will not recover from wilting is termed as permanent wilting point or wilting coefficient.

Chemical Properties: The chemical properties of soil can be studied under the following heads:

Soil Components: Soil contains a variety of elements, chemical compounds and minerals. Oxygen, alluminium, silicon, calcium, magnesium, sodium, potassium and iron chlorides, sulphates and oxides are usually present in the soil. Certain trace elements like cobalt, boron, iodine, cadmium, arsenic, zinc and barium etc. are also present in the soil. Certain organic compounds and dead animal or plant remains in all stages of decomposition are also present in the soil.

Percentage of Various Elements in Soil

Elements	Percentage	Elements	Percentage	
1. Oxygen	46.00%	6.	Sodium	3.00%
2. Silicon	28.00%	7.	Potassium	2.500%
3. Aluminium	8.00%	8.	Magnesium	2.00%
4. Iron	4 to 5.00%	9.	Trace elements	1.500%
5. Calcium	4.00%			

The chemical composition of soil is an important factor in determining the presence or absence of particular plants and animals. Peaty soils are not inhabited by animals for they are poorly aerated and more acidic.

Earthworms prefer soil rich in humus. Whether the soil is acidic or alkaline makes little difference to them. Similarly land snails are found in the soil rich in calcium salts which is required for their shell. Thus the soil considerably influences the distribution of organisms.

Soil pH: Soils of different places vary in their pH value. Soils with pH 7.0 are neutral while the pH above 7.0 indicates alkalinity and that below 7.0 indicates acidity.

Generally the pH value of soils lies between 2.2 and 9.6. Depending upon the pH value, soils may be neutral, acidic or alkaline. For example, the soils are alkaline in arid areas and acidic in rainy tropics.

The soil poor in calcium and magnesium are acidic in nature. Plants and animals in different soils vary with regards to their pH tolerance, *e.g.* pH acts as a limiting factor for the distribution of land snails as shown in the graph.

As evident from the graph neutral or slightly acidic soils are best suited for majority of the plants and animals. The highly alkaline soils with pH above 9.6 are usually devoid of plant life.

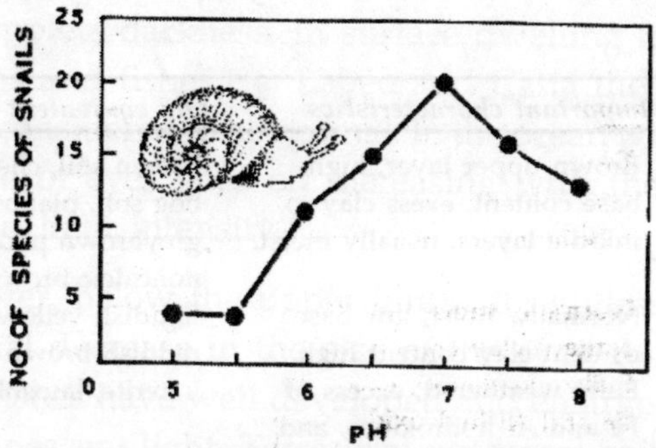

Graph depicting the effect of soil pH on the number of species of land snails.

Classification of Soils: Based on their physical, chemical and biological properties, soils are classified as follows:

1. Zonal soil classification
2. Comprehensive classification.

The zonal soil classification relates soils to the characteristics of the ecosystems in which they are formed but it is not precise for describing and classifying soils. The comprehensive system on the other hand stresses on the properties of the soils as they are found in the field with great precision in classifying them.

Comprehensive Soil Classification

Soil order	*Important characteristics*	*Old equivalent type*
1. Entisols	No stratification.	Azonal soils
2. Vertisols	Excess of clay.	Grummusols
3. Inceptisols	Moist, incomplete weathering, no clay, Fe and Al oxide saturated layers absent.	Humic clay soils, brown forest soils
4. Aridisols	Low organic matter, more than six months dry in a year.	Desert soils, sierozems, solonchak
5. Mollisols	Almost black, organic upper horizon, high base content.	Chestnut, chernozemes, rendzinas, prairie soils, brown forest soils, solonetz
6. Spodosols	Amorphous substances excessively present in middle horizons.	Podosol, brown podosol, ground water podosol

Contd....

Soil order	*Important characteristics*	*Old equivalent type*
7. Alfisols	Brown upper layer, high base content, exess clay to middle layers, usually moist.	Brown soil, chernozem, half bog soil, planosols, greybrown podosols, noncalcic brown soils
8. Ultisols	Normally moist, low base content clay content high.	Reddish yellow posol, reddish brown laterite
9. Oxisols	Fully weathered, excess of Fe and Al hydroxides, and clay in middle layers.	Laterite, latosols
10. Histosols	Excess of organic matter.	Bog soils

Source: Soil Survey Staff, Soil Classification, Soil Conservation Service (Washington, D.C. GPO, 1960).

Soil Types: Development and nature of soils are affected by several factors such as the nature of parent rocks, vegetation and climate. Calcium-rich rocks form soils rich in exchangable bases, siliceous rocks weather slowly forming shallow soils. Climatic factors such as precipitation and temperature influence the soil development as they affect physico-chemical processes, type of vegetation and the rate of decomposition of the organic matter.

The soils in wet regions are extensively leached, whereas the desert soils are not leached at all. The soils of cold regions are thinner and poor in iron as compared to soils of hot regions.

The type of vegetation also affects the soil type. The chemical composition of litter and its rate of decomposition determine the amount of organic matter, pH and base exchange capacity of the soils.

Following six major classes of soils with wide spread geographical distribution have been recognised:

1.	Soils of cold regions	:	Tundra soils
2.	Light coloured soils of temperate forest	:	Podzols

3.	Soils of the wet and subtropical forests	:	Latosols
4.	Soils between wet and dry regions	:	Prairie soils
5.	Soils of semi arid forests and grassland	:	Chernozems and chestnut soil
6.	Light coloured soils	:	Brown and desert soils

Soil Organisms: Depending upon the size, the vast majority of organisms, found in soil are classified into four major groups, namely microflora, microfauna, and microfauna. The meso and macroflora, no doubt, rooted in the soil, occur above the surface of soil and, therefore, are not included in the discussion.

Microflora: The microflora of soil comprises of bacteria, blue green algae, soil fungi and actinomycetes fungi.

(i) The soil bacteria are very useful and form the most numerous inhabitants of soil, constituting about 90% of the microflora. These are more abundant in neutral soil rich in organic nutrients. These are heterotrophic bacteria and autotrophic bacteria.

(ii) The heterotrophic bacteria derive their nutrients from organic matter of the soil. These bring about the decay and fragmentation of dead and decaying organic matter, thereby releasing considerable amount of nitrogen, phosphorous and other nutrients. These substances increase the fertility of soil.

(iii) The autotrophic bacteria are nitrogen fixing bacteria (nitrifying bacteria), sulfur bacteria, iron bacteria, manganese bacteria and carbon dioxide bacteria etc. These synthesize organic substances rich in nitrogen or other minerals.

(iv) The soil fungi are found in acidic soil. These may be parasitic, symbiotic or saprophytic. The parasitic fungi cause several plant diseases or produce toxic substances.

The saprophytic fungi release simple nutrients by the breakdown of dead and decaying organisms. The symbiotic fungi live in the roots of certain plants and both fungus and plants are benefitted.

(v) Actinomycetes fungi are found in saline soils and bring about decomposition of organic matter. These produce numerous important antibiotics.

Microfauna: It includes animals with body size within the range of 20m-200m. These include protozoans, mites, nematodes, rotifers, tardigrades, etc.

(i) *Protozoans:* Most common protozoans inhabiting the soil are the flagellates, amoebae and ciliates occurring near the surface soils, while the forms like Euglypha, Difflugia and Thecamoeba have vertical distribution.

(ii) *Nematodes:* There are the most abundant and widespread soil organisms comprising 80-90% of all multicellular animals. These abound by as much as 20 million per square metre in grassland to 30 million per square metre in raw humus soils. The well known examples of plant parasitic nematodes are 'root knot' or 'gall forming' nematodes like Meloidogyne and cyst nematodes like Heterodera. Mononchus is a predatory nematode which occurs in almost all soils preying upon other nematodes and micro-organisms.

Mesofauna: This group includes micro-arthropods and insect larvae. These organisms play a great role in maintaining the soil fertility. Their size varies from 200μ to 1cm.

(i) *Collembola:* Collembolans are springtails forming one of the predominant group of soil mesofauna. Common soil collembolans are Isotomids, Onchiurids, Entomobryids and Podurids. Some of them are phytophagus and are serious crop pests, *e.g.* Sminthurus viridis (pest of clover and some grasses), Bourlertiella hortensis (pest of peas, beet and potatoes). Collembolans mostly feed on fungi, bacteria and nematodes.

The isopterans (termites) like Reticulotermes and Odontotermes are important soil-dwellers and play an important role in the break up of organic materials and their mixing up with mineral soils.

(ii) *Acarina (mites):* They also form a important group of soil organisms. Mites belonging to Cryptostigmata are herbivorous and feed on rotting plant tissues, fungi, algae and lichens. These are most abundant in forest litter and grass meadows. Since Acarina include predatory forms feeding on collembolans, pauropods, nematodes, etc., fungi and virus and bark feeding forms, they are important soil mesofauna. They greatly influence on soil texture, organic matter content, vegetation, microclimate, water and other soil flora and fauna.

Other important mesofauna organisms are some myriapods like Pauropus and Eurypauropus, they feed on decaying plant material and fungi. Centipede-like myriapods are also of great importance as they increase the fertility of soil.

Macrofauna: It includes those animals whose body size is greater than 1 cm. It includes earthworms, centipedes, millipedes, pseudoscorpions, wood lice and insects belonging to different orders such as Psocoptera, Embioptera, Zoraptera, Hemiptera and grasshoppers. The burrowing annelids include Lumbricus, Glycera and Arenicola. Certain ants and insects larvae are also quite common. Termites also form an important group. Among vertebrates adapted for burrowing life are Ichthyophis, Cacopus, Sytema, Breviceps (Amphiba), Sphenodon, Uromostix, limbless lizards and snakes (Reptiles), Talpa, Dasyurus, Notroryctes and among mammals, the insectivores and rodents.

5

Role of Temperature

Temperature is one of the most common ecological factor. It has an universal influence and is frequently a limiting factor for the growth and distribution of animals and plants. It also controls reproduction, rate of embryonic development, migration and a number of behavioural characteristics of the organisms.

Temperature Variants

The range of temperature varies greatly in different environments such as freshwater bodies, marine and terrestrial. In the open water of ponds and lakes, temperature never falls below 0° and in oceans never falls below-2.5°C. The maximum temperature recorded in marine water has been 36°C in Persian Gulf. On land the minimum temperature of-70°C has been recorded in Siberia and the highest temperature of 60°C in desert areas. The temperature of hot springs may reach upto 100°C.

Variation in Temperature: Generally the temperature of the air near the land surface is 17°C higher in the day time than at night. In deserts this difference may be as much as 40°C. But in water bodies like a deep lake, the difference between day and night temperature is usually less than 1°C. The maximum diurnal change in temperature in marine environment is about 4°C. With increasing depth, the amplitude of day and night change in temperature is reduced and no change is evident below a depth of 15 meters.

Thermal Stratification

***Changes in Temperature*:** Thermal stratification or seasonal changes in temperature can be best studied in large deep lakes of the temperature region. Fresh-water has maximum density at 4°C. Thus any warmer or colder water will float on top of water of this temperature. During winter, the temperature falls and the surface water is frozen. This ice prevents the cooling of water below it. With the onset of spring the ice melts till temperature rises upto 4°C and entire mass of water has the same uniform temperature.

In summer, the atmospheric temperature may rise to about 25°C. This warms the upper surface layers of water to about 21-12°C. The upper, warm layers being lighter remain at the top while the temperature of bottom layers is hardly 5°C. In between the two limits lies an intermediate zone with varying temperature from 9°C-21°C. This is the thermocline. This is the zone of rapid vertical temperature change with the thermal gradient of 1°C per meter. The upper warm layers are referred to as epilimnion and the bottom cold layers are known as hypolimnion. Similar seasonal cycle of temperature occurs in the ocean but it depends on the location and the season.

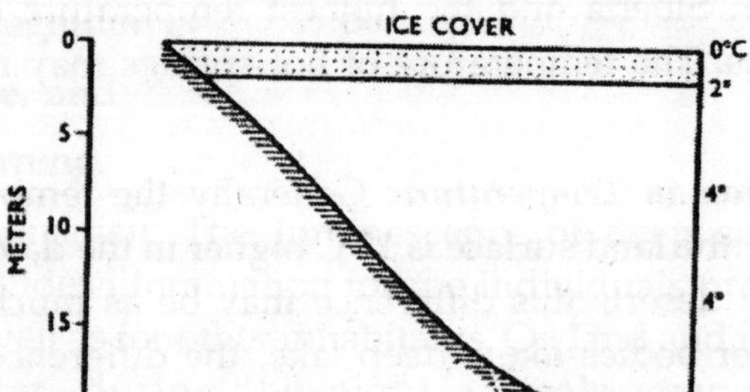

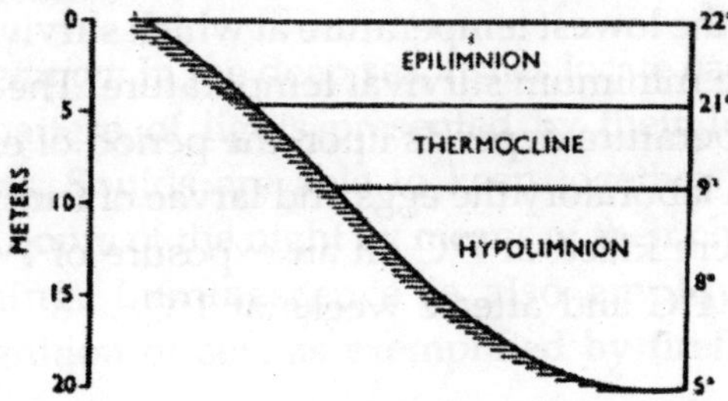

Thermal stratification in a deep lake in temperature region during winter and summer (Redrawn from Concept of Ecology by Kormandy).

The temperature changes in the lakes and ocean are much less and much slower than in the terrestrial environment. Further, organisms can easily get out of excessively high or low temperatures by a short journey into deeper water. Thus life in water is much easier than life on land. But, whenever an unusual change in temperature occurs, dire results may follow with high mortality of the aquatic organisms.

Temperature Tolerance: Every organism has a range of temperature that it can tolerate. Based upon temperature tolerance, organisms are classified as eurythermal and stenothermal. Animals which can tolerate a wide range of temperature are referred to as eurythermal, *e.g.* cyclops, oysters, toads wall lizard, sperm-whales, man etc. On the other hand animals with narrow range of temperature are referred to as stenothermal, *e.g.* snails, corals, termites, fishes and most reptiles.

The temperature at which the activities of an organism are at the maximum rate is known as optimum temperature. There are maximum and minimum limits of temperature above and below which organisms of a species could not survive.

Minimum Temperature: The lowest temperature at which an organism can live indefinitely in an active state is termed as the minimum effective temperature. With a further reduction of temperature, the organism goes into a condition of inactiveness called the chill coma. The activities of the animal can be revived by raising the temperature to minimum effective limit. Thus the lowest temperature at which survival is possible is called the minimum survival temperature. The actual value of this temperature depends upon the period of exposure. For example, in laboratory the eggs and larvae of fruit fly, Creatitis capitata, were killed at 7°C on an exposure of 7 weeks, after 3 weeks at 4°C and after 2 weeks at 1°C.

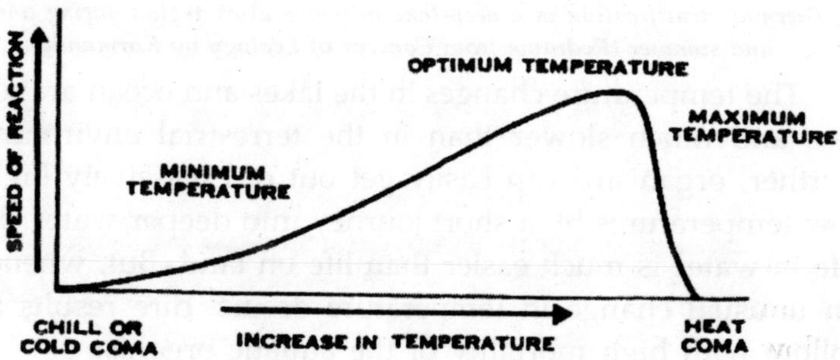

Effect of temperature on the physiological activity of animals.

Maximum Temperature: The highest temperature at which the organism can live indefinitely in an active state is referred to as maximum effective temperature. At higher temperatures the organism goes into heat coma but recovers when restored immediately to cooler conditions.

Generally the range of temperature tolerance within which the living organisms could carry on their life activities lies between 10-48°C. But certain organisms could survive and carry on their metabolic activities at as low as 0°C or as high as 88°C. At low temperature tissue damage and death due to freezing is caused by the formation of ice crystals. At high temperature death occurs on account of thermal denaturation of intracellular enzymes and other cellular proteins.

The favourable temperature range for any particular species is the temperature at which normal physiological activities of the animal takes place. This can be exemplified by the rotifer Keratella procurya (NAYAR, 1971). This species appears in the ponds of pilani, Rajasthan when the temperature is below 24°C and disappears when it rises above 24°C. Its distribution reaches its peak during the months of October to March, when there is a fall in temperature.

Since there is no great difference between maximum and minimum temperatures in water aquatic animals possess a narrow range of temperature tolerance. Exposure to temperatures beyond the favourable range may result in the death of organisms or it may retard their growth and reproduction.

Let us now consider the cases of high temperature and low temperature that could be lethal to organisms. Many plants and animals are killed by temperatures that are too low for them but are nevertheless far above values at which tissues would actually freeze. Similarly, at the other end of the scale there are many instances of death from excessively high temperatures but at values below those producing heat coagulation. Lethal extremes vary greatly from species to species. An organism may be killed by a degree of chilling which is insufficient to cause direct damage to the tissues. As a matter of fact no animal or plant tissues freeze at 0°C. They undergo freeing usually at much lower temperatures. It is a matter of common occurrence that animals and plants die even before their surrounding temperatures have reached 0°C. If the tissues do not freeze, why are these animals and plants killed by moderately low temperatures. This is because a negative correlation exists between low temperatures and chemical reactions.

Most chemical reactions are slowed down by lowering temperature and eventually stop. The stopping of anyone of the chemical reactions or vital process is sufficient to cause the death of the organism. However, various chemical reactions going in the body come to a stop at different temperatures. Under certain circumstances the organism may even die when all the processes are still going on. It may be due to the fact that the different reactions are slowed down by different amounts with the result that mutually dependent processes get out of tune and cause the death of the organism.

The cause of death of organisms at higher temperature may be due to desiccation and increased rate of metabolism. There is an optimum temperature for every vital process. When the temperature exceeds this optimum, the reactions get out of adjustment and death results.

The rate of chemical reactions increases with the rise in temperature. This has been expressed by Van't Hoff's rule according to which the rate of chemical reactions is doubled for every 10°C rise in temperature. This increase is represented by Q_{10}. In other words Q_{10} represents the effect of temperature on reaction rate, and is in the range of 1.5 to 3 for most metabolic processes. Q_{10} for a particular animal varies with the habitat temperature to which it is adapted.

This lays emphasis on the importance of the choice of temperature range. As a matter of fact the differences in oxygen consumption at comparable stages of development are purely a function of temperature. For example, during development in the lygaeid bug, the Q_{10} for oxygen consumption is 3-3.8 which drops to 2.5 after the larvae hatch (Glenn Richards, 1964).

Temperature plays an important role in the life of the bed bug (Johnson,1940, 42). The number of generations formed during a year directly depends on the ambient temperature. Thus 12 generations in the tropics, and two in cold climates have been observed. The resulting influence of temperature on the death rate of eggs of Cimex when exposed to three different temperatures for varying periods is shown in figure.

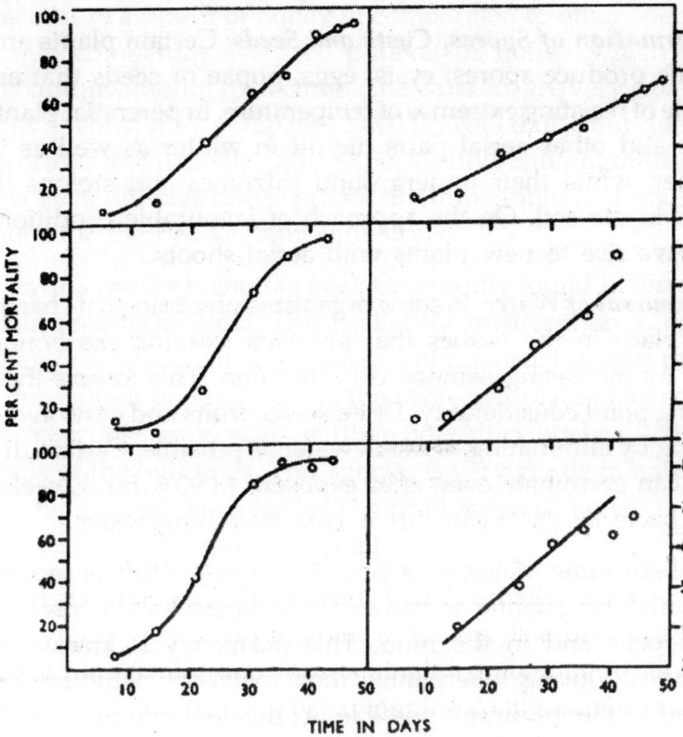

Diagrammatic representation of temperature on the death rate of eggs of Climax. Bottom temperature -42°C, middle -9.8°C, top -11.7°C, Left Death rate percent plotted against time; Right: Probit.

Poikilothermic Organisms: Animals are broadly classified into homoiothennic (warmblooded) and poikilothermic (cold-blooded) animals. In cold-blooded animals body temperature fluctuates with that of the environment, *e.g.* fishes, amphibians and reptiles. The warmblooded animals are those animals which regulate their body temperature at a constant level irrespective of the environmental temperature, *e.g.* birds and mammals.

Meeting Temperature Extremes: The animals and plants have developed certain special mechanisms by which they tend to overcome the harmful effects of temperature. Some of them have been enumerated below:

Formation of Spores, Cysts and Seeds: Certain plants and animals produce spores, cysts, eggs, pupae or seeds that are capable of resisting extremes of temperature. In perennial plants leaves and other aerial parts die off in winter as well as in summer, while their underground rhizomes and stolons lie buried in the soil. On the approach of favourable conditions they give rise to new plants with aerial shoots.

Removal of Water: In some organisms physiological change takes place in the tissues that prevents freezing the bound water by increasing osmotic concentration. This lowers their freezing point considerably. Dried seeds, fruits and cysts avoid freezing by eliminating as much water as possible. Certain dry seeds can germinate even after exposure to 90°C for 3 weeks. Some bacterial cysts can thrive even in boiling water.

Hibernation: Majority of the cold-blooded animals remain dormant during winter season and hide themselves in crevices, under-rocks and in the mud. This dormancy is known as *hibernation.* During hibernation, their metabolic activities are reduced to the minimum possible. At this low rate of activity, little energy is drawn from the body reserves such as fats and glycogen. Thus hibernation is characterised by reduced metabolic rate, low body temperature and reduced rate of heart beat.

Toads and frogs hibernate in solitary burrows on land. Reptiles aggregate in large numbers below rocks or burrows. Aquatic turtles tide over winter buried in mud below the ponds. The common mammals that hibernate are the monotremes, insectivores, bats, some carnivores and rodents. Bats (microchiroptera) are very good examples for hibernators.

Animals like badgers, bears and raccoons also exhibit dormancy in winter. This cannot be called as hibernation as there is no fall in their body temperature during dormancy and this takes place at intervals. This may better be called as pseudohibernation or carnivorian lethargy.

The factors that bring about hibernation in animals are the following:

(a) gradually decreasing temperature

(b) lack of food and starvation

(c) inability to move about due to snow covered land

(d) dry food

(e) glandular disturbances

(f) accumulation of fat

(g) lack of proper heat regulating mechanism.

The period of hibernation depends upon the length of the winter. Though some animals hibernate during the whole winter others do so only during the severe months. Hibernation occurs in temperate, arctic and antarctic regions and not in tropical countries.

Aestivation: Aestivation is the period of dormancy that is exhibited by animals during summer. This is commonly met with in insects, some invertebrates and vertebrates. According to Hora some snails remain quiescent during continuous summer rains. The lung fishes spend their summer inside mud cocoons when the ponds go dry and come out only when the rains come.

The Ophiocephalids of the Indo-Malay region remain buried in the rice fields during summer. Terrestrial molluscs like Ariophanta seek cool and shady places during summer and remain dormant by closing the aperture of their shells with a thin membrane, the epiphragm. Among reptiles snakes, lizards and tortoises are good aestivators.

Some Australian frogs become turgid and swollen with water and bury themselves to aestivate till the next rain. Dragon fly nymphs were noticed by Tillyard to aestivate for months. The grand squirrel of California is a good aestivator. The lung fish, *Protopterus,* in order to escape summer extremes, burrows into the mud and secretes a cocoon of slime around itself. It does not feed but subsists on the reserve body fats.

Insects exhibit dormancy in the form of diapause. Diapause is a stage in the life of an organism during which growth and development are suspended or slowed. The groundnut pest, Amsacta albistriga remians dormant (diapause) in the pupal stage for nearly eight months.

Changes during Hibernation and Aestivation: In dormant animals the rate of metabolism is very slow. Though, the frogs loose fat and sugar during hibernation, the development of their gonads proceeds through slowly. Warm blooded animals temporarily become cold-blooded during hibernation. There is a drop in body temperature in dormant mammals. This may take place in one stroke as in hamster and ground squirrel. Dormant mammals often show heart block when the heart beat rate may be reduced to two or three per minute.

The respiratory movements are slow, and the pulse rate comes down. Hibernating insects can survive lower temperature than what normal insects can tolerate. Dormant animals are usually resistant to parasites. The calcium content of the molluscs is reduced during hibernation. The activities of the important endocrine glands like pituitary, thyroid and adrenals are retarded. Animals develop special insulating materials like fur, feather and fat during hibernation. Aestivating animals develop a hard covering like cocoon, cyst etc. to protect themselves from drying.

Arousal from hibernation is explosive and certainly the most dramatic aspect of the hibernation cycle. As the animal starts to come out of sleep, its body temperature rises rapidly, perhaps from 4°C to 17.5°C in an hour and half (MAYER, 1960). In ground squirrel, when the anterior part of the body has reached about 36.5°C, the temperature of the hind parts rises rapidly, due to dilation of the blood vessel. The animal's hair become erect and its body shakes. At 17.5°C the shivering stops and the squirrel moves its tail. At 24°C the animal opens its eyes and suddenly sits up. At this time the temperature rises rapidly again. When the body temperature reaches 24 to 25.4°C, about three hours from the time of arousal, the ground squirrel is warm and active again.

Warm-bloodedness: This is a special device to meet temperature extremes and remain in active condition. Birds and mammals are able to maintain a constant body temperature despite variations in the environmental temperature. In summer the body temperature is brought down to normal by allowing regulated water evaporation from their body surface. Evaporation brings cooling effect to the surface. During cold weather, birds and mammals maintain their normal body temperature by the insulation action of feathers, fur and fat and also by suitable physiological adjustments.

Thermal Migrations: To escape the extremes of temperature certain animals migrate to places where they can have optimum temperature. Such movements are known as thermal migration. Such migrations may be from centimeters to several hundred miles. Thermal migrations include movements from exposed scorching heat to shade in deserts and from shade to sun in cold regions.

Many desert animals have become nocturnal in habit and thus avoid the heat of the day. Frogs, turtles and other amphibians make short trips into or out of water. Burrowing animals escape excessive heat or cold by going deeper into the soil.

With the advent of winter the bear, deer and other game animals descend from the mountains into the valleys. In spring they again return to higher elevation. Many fishes and other aquatic animals leave the shore in summer when the water has become too warm. Insects, birds and mammals undertake long distance migration to escape from climatic extremes. The migration habits of birds and other animals depends not only on temperature but several other factors such as food, breeding and day length.

Effects of Temperature: Temperature has manifold effects on the structure, physiological processes, behaviour and distribution of most of the organisms. Some of these effects are as follows:

Metabolism Effects

Temperature has direct effect on the metabolic activity of the organisms. Most of these activities are under the control of enzymes which in turn are influenced by temperature. Initially with the increase in temperature, enzymatic activities also increase, thereby increase the rate of metabolism. For example, the activity of enzyme liver arginase goes on increasing with the increase in temperature from 17°C to 48°C. But any increase in temperature beyond this limit brings about retardation in the activity of the enzymes. Similarly chirping of crickets is higher in warm weather and lower in cool weather.

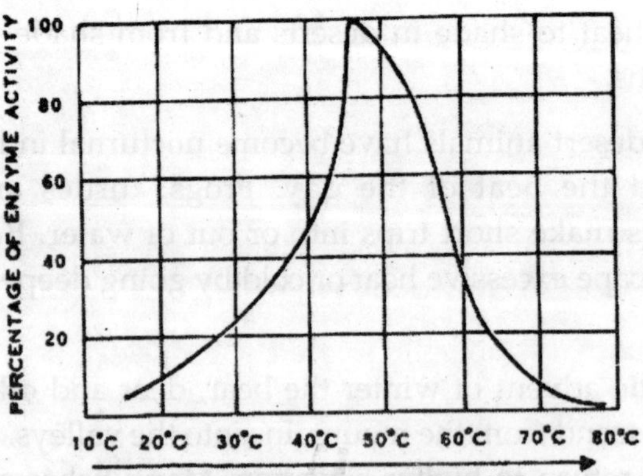

Effect of temperature on the activity of liver enzyme arginase.

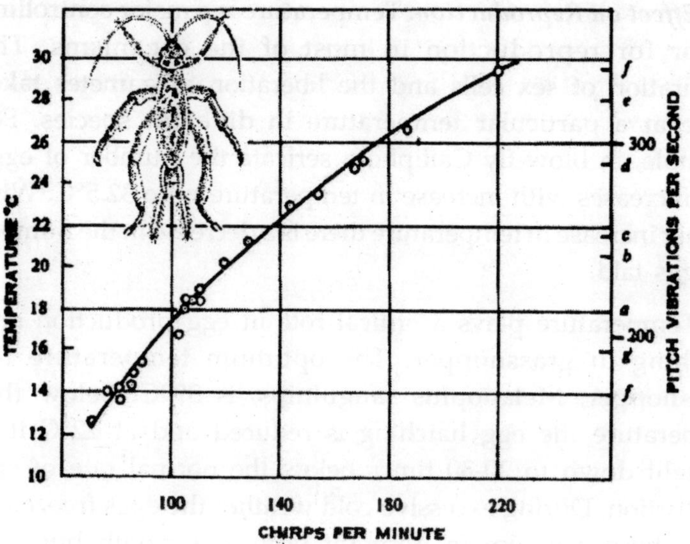

Relation of chirping rate and temperature in tree cricket Oecanthus.

Effect on Reproductive Behaviour: Reproductive behaviour in a large number of animals have been found to be governed by temperature. Some animals breed uniformly throughout the year, some mainly in summer, others exclusively in winter white still others have two breeding periods, one in spring and the other in the fall.

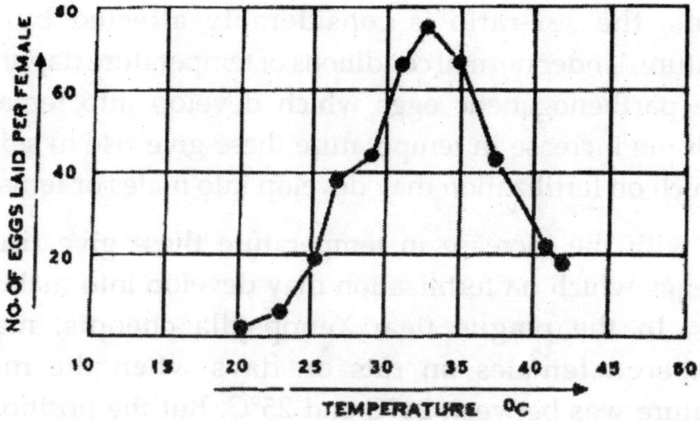

Effect of temperature on the number of eggs laid per female in blow-fly.

Effect on Reproduction: Temperature is a major controlling factor for reproduction in most of the organisms. The maturation of sex cells and the liberation of gametes takes place at a particular temperature in different species. For example, in blow-fly Calliphora sericate the number of eggs laid increases with increase in temperature upto 32.5°C. With further increase in temperature there is a decrease in the number of eggs laid.

Temperature plays a critical role in egg production and hatching in grasshopper. The optimum temperature for grasshopper, Melanoplus sanguinips, is 39°C. Below this temperature the egg hatching is reduced and at 22°C it is brought down to 20-30 times below the normal rate of egg production. During excessive cold weather the eggs freeze. At 23°C these can remain dormant upto one month but at a temperature of 29°C, these are unable to survive.

The egg laying and hatching is also associated with drought. The grasshopper outbreaks usually coincide with the extended periods of hot and dry weather. But the severe drought also reduces the fecundity of females because of reduction in the succulent food.

Effect on Sex-ratio: In certain animals such as rotifers and daphnids, the sex-ratio is considerably affected by the temperature. Under normal conditions of temperature daphnids produce parthenogenetic eggs which develop into females. But with the increase in temperature these give rise to sexual eggs which on fertilization may develop into males or females.

But with the increase in temperature these give rise to sexual eggs which on fertilization may develop into males or females. In the plague flea, Xenopsylla cheopis, males outnumbered females on rats on days when the mean temperature was between 21°C and 25°C. But the position is reversed on more cooler days.

Effect on Growth and Development: Temperature also affects the growth and development considerably at different stages in the life-cycle of an organism. The optimum temperature for the development of trout eggs is 8°C. The adult trout does not grow in water with a temperature of less than 10°C and the maximum growth takes place between 13-19°C. In cold, egg development starts at 1°C. It increases regularly upto a temperature of 14°C and above it the egg does not survive. Sea urchin, Echinus esculentus, attains the maximum size in warmer waters.

Mackerel egg starts developing at 10°C. Form 10°C to 21°C, the rate of development increases and at the upper value, the egg will hatch in just 50 hrs. AT 25°C no development takes place. The effect of temperature on the growth of Mackerel eggs can be best understood as shown in the table.

Relation between Temperature and Rate of Development of Mackerel eggs:

S.No.	*Temp.*	*Development*	*S.No.*	*Temp.*	*Development*
1	8°C	No development	5	18°C	Eggs hatch in 70 hrs.
2	10°C	Eggs hatch in 207 hrs.	6	20°C	Eggs hatch in 60 hrs.
3	12°C	Eggs hatch in 150 hrs.	7	21°C	Eggs hatch in 50 hrs.
4	25°C	Eggs hatch in 70 hrs.	8	25°C	No development

(From Elements of Ecology by Clarke)

The time required for the development at various temperatures is also an important feature in the three species of malarial parasite in their:

Species	*16°C*	*17°C*	*18°C*	*20°C*	*24 °C*	*30°C*	*36°C*
1. P. viva	–	halted	–	16 days	9 days	7 days	None developed
2. P. falciparam	–	–	halted	20 days	11 days	9 days	None developed
3. P. malariae	halted	–	–	30 days	21 days	19 days	None

(After Garnthm, 1964)

Bhatia and Kaul (1966) have made elaborate studies on the effect of temperature on the development and oviposition of the red cotton bug Dysdarcus koenegii.

According to them the lower and upper limits for the development of eggs seems to be between 12.5 and 1°C; and 32.5°C and 35°C respectively. With the rise of temperature from 15 to 32.5°C, the duration of incubation period decreases and the viability of the eggs is lower at the extremes of temperature.

The nymphs have a narrower range of effective temperature than the eggs (0-35°C) and the duration of the life-cycle at 20, 25, 27.5 and 30°C reduces gradually. With the decrease in the temperature from 35°C to 15°C, there is lengthening of the premating and preoviposition periods.

Effect on Animal Distribution: Extremes of temperature as well as varying period of exposure to heat and cold kill many animals. This is the main reason why temperature acts as a limiting factor and affects the special distribution of animals.

Regions of very low and high temperature exclude many species of animals. Temperature acts as a decisive factor in the distribution of animals in the following ways:

(i) Temperature along with air currents and moisture constitute climate and this influences animals in a composite way.

(ii) Availability of particular types of food in sufficient quantities determines the distribution of many animals and heat affected the distribution of food animals and food plants.

(iii) Thermal stratification brings about thermal zones which vary from two to thirty degree centigrade. The thermoclines between the zones act as barriers for the stenothermal animals.

(iv) Heat alters the density and viscosity of water and thereby reduces buoyancy and this affects the floating ability of many large aquatic animals.

(v) Temperature restricts benthic form to particular coasts. The eastern and western coasts of Africa and India have different benthic animals.

(vi) Animals require for their existence limiting minimal temperature in winter and necessary maximal temperature in summer.

Both these factors together determine the boundaries of distribution.

(vii) The early developmental stages of most of the organisms are vulnerable to temperature changes and this brings about heavy mortality and thereby affect distribution.

(viii) In many animals low temperature prolongs the larval life exposing them to the rigours of life for a long time and thereby the chances of survival are very much reduced.

(ix) A relatively constant temperature or slight variations in temperature always attracts many animals and leads to the concentration of different species.

(x) Rigid stenothermal animals like oysters, lobsters etc. have restricted distribution due to the occurrence of optimal temperature conditions only in some patches in the sea.

(xi) The desert mammals and insects which aestivate during the summer will not reproduce unless they undergo a spell of cooler climate through some part of the year. The gemmules of sponges hatch out after a period of cooling and the eggs of Australian grasshoppers will not grow after a particular stage unless they are exposed to low temperature.

(xi) Bipolar distribution is a standing example of the effect of temperature on distribution.

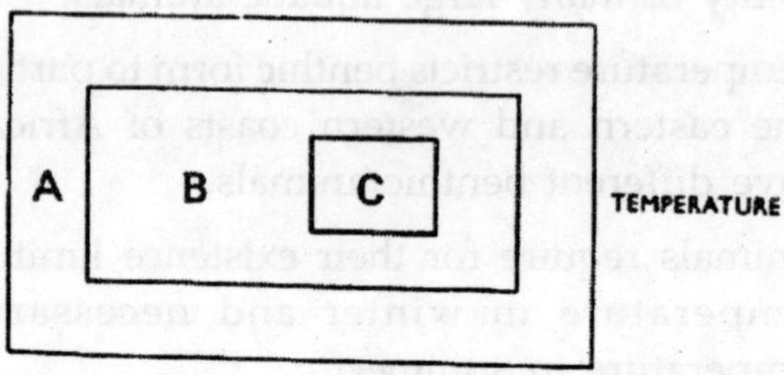

Diagrammatic representation of 3 tier distribution of animals.

In general, animals have a three tier distribution according to their living conditions, (A) an area where they have optimum conditions for life, (B) an area within A where they become sexually mature, and (C) an area in B where they can reproduce.

The distribution of eurythermal animals is not generally much affected by any variation of temperature.

Temperature and Moisture: The interaction of temperature and moisture depends on the relative as well as the absolute values of each factor. Thus temperature exerts a more severe and limiting effect on organisms when moisture conditions are either very high or low, then when such conditions are moderate. Similarly moisture plays a more critical role in the extremes of temperature when the humidity is low or moderate than when it is very high. Conversely, hot humid weather is less favourable for the weevil.

Structural and Behavioural Effect: Several structural modifications and behaviour patterns are induced by temperature. Animals living in colder regions have a greater longevity than the forms living in warmer zones and are larger in size.

In Drosophila melonogaster, temperature affects the position of genes and chromosomes behaviour during crossing over. This brings about the change in the number of facets in the eyes, size of the vestigial eyes and the presence or absence of more or less eggs than the normal number of six.

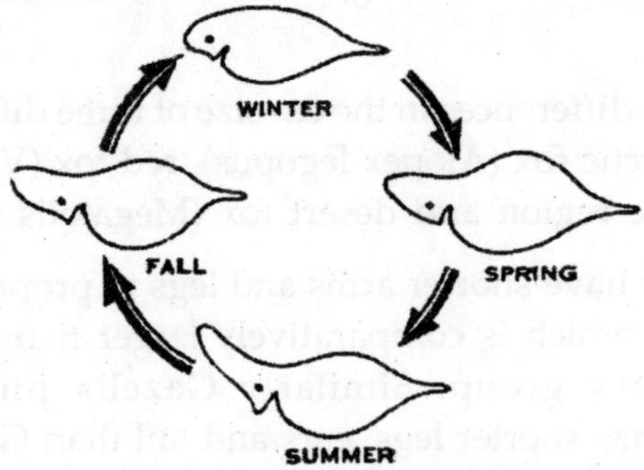

Cyclomorphosis in Daphnia.

Cyclomorphosis: In certain planktonic animals, the body form changes with the seasonal changes in temperature. This phenomenon is known as cyclomorphosis. The process of cyclomorphosis was first described by Coker (1939) in certain cladocerans (*e.g.* Daphnia).

In winter cladocerans have a round head. During spring a helmetlike projection develops on the head. In summer, this projection attains maximum size and in winter the head again becomes rounded. These prolongations of the helmet have been interpreted as an adaptation aiding floatation since the buoyancy of water becomes reduced as the temperature increases (Fig.),

Gloger's Rule: Temperature along with moisture and light affects the colouration of some animals. In warm humid climates many mammals, birds and insects are darker in colour than their counterparts living in cool or dry climates.

Jordan's Rule: Temperature has an apparent control on the number of vertebrae in certain species of fish. Thus fishes of low temperature waters have more vertebrae than those of warm-water forms *e.g.* cold-fish.

Allen's Rule: Tail, ears, bill, neck and other extremities of animals are relatively shorter in the cooler parts than in the warmer parts and are in general characterised by compact structure.

It shows differences in the ear size of three different species of fox, an arctic fox (Alopex legopus), red fox (Vulpes vulpes) of temperate region and desert fox (Megalotis zerda).

Eskimos have shorter arms and legs in proportion to their trunks size, which is comparatively larger than in any other contemporary group. Similarly Gazella pincticanda of Himalayas has shorter legs, ears and tail than Gazella benetti found in the plains of Himalayas, though both are of the same body size.

Allen's rule has been tested in laboratory and holds true for species reared in laboratory. It is seen that tails of individuals reared in a temperature range of 15.5°C to 20°C are noticeably shorter than those that are reared in a temperature range of 31°C to 33.5°C, even though these are of the same species.

The adaptive significance of the feature is obvious that short extremities reduce exposed surface which reduces loss of heat from the body.

Bergman's Rule: According to this rule, geographic races of a species inhabiting cold regions are much larger than the forms inhabiting the warmer regions. For example, the birds and mammals in the colder regions are much larger than the related forms inhabiting the warmer regions.

Bergman's rule suggested that as the temperature decreases from the equator towards the poles, the body size of the individuals living there increases gradually. The significance, of the increases in size is explained by presuming that a larger animal has less surface area per unit of weight than the smaller one and thus proportionately less heat is lost by radiation than in a smaller animal. In warmer zones on the other hand the large size will be harmful to these animals, hence they are comparatively short statured. The rule can be supported by facts obtained by the study of nature. The largest bears are the polar bears and the Rodiak bears which are found in the far north while the smaller black bears are found in more temperate regions.

Similarly penguins which are found in Antarctica attain a body length of 1000 to 2000 mm. whereas penguines of the equatorial Galapagos islands are about 490 mm long.

Role of Light

Light is an important ecological factor which influences the growth and distribution of plants and animals. Light is indispensable for photosynthesis in plants. The life-giving sun's energy is trapped in the carbonic food by green plants. It is responsible for directing and controlling the structural and behavioural characteristics. Light also acts as a limiting factor in the many sided activities of the organisms.

Spectra: The spectral distribution of solar energy that reaches the earth's surface. The solar energy consists of electromagnetic waves composed of high frequency and low frequency wavelengths. In general, high frequency wavelengths from 3600Å (360 millimicrons) downwards are ultraviolet rays, X-rays, gamma rays and cosmic rays and low frequency wavelengths from 7600Å (7.0 millimicrons) upwards are infrared, radar, microwaves and radiowaves.

The part of the solar energy that can be perceived by the human eye is called the visible light. The visible light is made up of a series of colours ranging from violet to red through indigo, blue, green, yellow and orange. The whole range of colours constitutes the visible spectrum. The frequency of wavelengths in visible light ranges from 3600Å 7600Å. In the figure of spectral distribution of the solar energy, about one-half of the total solar radiation is in the infra-region, visible light also constitutes about one-half of the total solar radiation while the ultra-violet light constitutes only a small fraction of the total solar radiation.

Light on Land: The intensity of light reaching the earth's surface depends on two factors-(1) The angle of incidence and (2) the absorption of light rays by the different layers of the atmosphere. The lower the altitude of the sun, the smaller is the angle of incidence but longer is the path of light through the atmosphere with corresponding reduction in intensity. The greatest intensity of sunlight occurs when the sun is overhead. At higher altitudes the intensity of light is correspondingly reduced. The intensity of light is affected by the presence of clouds, moisture and dust in the atmosphere.

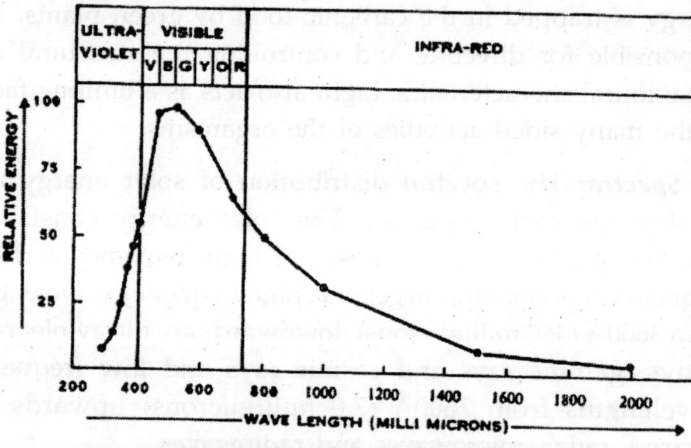

Spectrum of tadiant radiant energy and visible light.

In dense forests, the intensity of light on the forest floor is greatly reduced due to the luxuriant growth of the forest vegetation.

The total amount of light received by an organism is determined by the intensity and duration of light. The duration of light for which an organism is exposed to light is called photoperiodism. At the equator, light duration is of 12 hours.

In the temperate regions the length of the day increases with the approach of the spring season, while in the polar regions the day becomes 24 hours long during summer.

Light in Water: About 10% of the total light falling on the water surface is reflected back. The remaining light while passing downward in water is modified in respect of intensity, spectral composition and distribution etc. The suspended particles in water reduce the light intensity both by absorbing and scattering the light.

The rate of reduction in light intensity is known as the extinction rate. Pure water causes the extinction of light at different rates in different parts of the spectrum. Red rays are completely absorbed at a depth of 4 meters. Orange rays are absorbed at a depth of 20 meters, while the yellow rays penetrate upto 80 meters and green rays upto 80-100 meters. At depth of 100 meters car more the blue light becomes completely predominating. Below a depth of 200 meters or so lies the zone of perpetual darkness.

Percentage of Incident Light

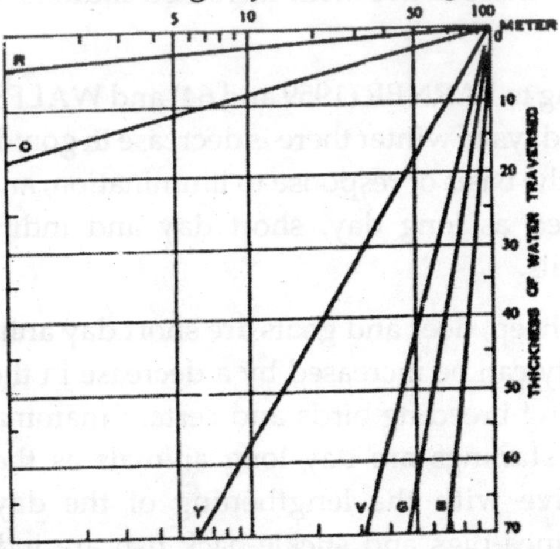

Reduction in the intensity (logarithmic scale) of the colour components of sun light (indicated by initial letters) at increasing deoths (linear scale) in optically pure water.

Biological Effects of Light

Effect on Metabolism: The absorption of light rays of the visible part of spectrum has little effect on organisms. The violet and ultraviolet rays are harmful and produce photochemical changes in the organisms and, therefore, influence their metabolism.

The intensity of solar radiation affects living organisms at different latitudes and seasons of the year. Within optimum limit, photosynthesis increases in direct proportion to the intensity of light. Cold blooded animals aestivate or hibernate as the solar radiation increases or decreases. Unicellular organisms including bacteria, algae, protozoa and eggs of vertebrates and invertebrates are killed by exposure to ultraviolet rays.

Effect on Reproduction: In many animals light initiates the breeding activities by stimulating their gonads. There appears to be a definite relationship between length of the day and egg laying in birds. ROWAN (1931) reported that gonads of some birds become more active with increased illumination during summer.

According to FARNER (1959 and 64) and WALFSON (1960) during short days of winter there is decrease in gonidial activity of birds. On the basis of response to illumination, animals have been classified as long day, short day and indifferent day length animals.

Certain sheep, deer and goats are short day animals. Their sexual activity can be increased by a decrease in the length of the day. Spring breeding birds and certain mammals such as turkeys and starlings are day long animals as they become sexually active with the lengthening of the day. Ground squirrels, guinea-pigs and stickle-back fish are indifferent to day length, as they are least affected by the short or long day periods.

Developmental Factors

Since light influences metabolism, it affects the growth and development of organisms. For example, Salmon larvae undergo normal development only when sufficient light is present. In the absence of light their development is not normal and there is heavy mortality. Mytilus lavae in their earlier stages grow larger in darkness than in light.

Effect on Pigmentation: Light induces certain chemical changes which result in the formation of photoreceptors in the form of pigment spots. It may influence the pigmentation of animals in the following ways:

(a) *Skin Colour*: Characteristic lack of pigment in cave animals is associated with darkness (*i.e.* total absence of light). Certain aquatic animals loose their colour when shielded from light. RASQUIN (1947) showed that cave amphibians (*e.g.* Proteus) and fishes with little or no colour, when exposed to normal light, developed abundant pigment in the skin. The darkly pigmented skins of human inhabitants of the tropics also indicate the same effect.

(b) *Protective Colouration*: The pigmentation of a number of animals acquires a colouration that can protect them from enemies. Such colouration is known as protective colouration. One common type of protective colouration is a simple matching of body colour in respect and pattern to the background. For example, colouration in case of a quail squatting in the grass, moth on the bark of a tree and leaf-insect (Phyllium) among green leaves is exactly like that of the background. It is exceedingly difficult to distinguish them from their surroundings. The arctic hare, weasel and ptarmigan show seasonal colour changes from brown in summer to white in winter. This is clearly related to the conspicuousness of such animals against bare ground or snow-covered landscape.

A second type of protective colouration is obliterative shading in which the birds, mammals or fish display darker colour on the back and lighter colour underneath. This difference counteracts the stronger illumination received from above and the animal blends with its background.

(c) *Colour Changes*: Certain animals have the capability to change their colour according to their surroundings. Frogs and chameleons are well known examples. According to PROSSER and BROWN (1950) colour changes are brought about by visual stimulus. Visual stimulus is the ability of the animals to change the colour to suit the environment in which they live. Colour changes are widely found among crustaceans, insects, fishes, amphibians and reptiles. These help the animals in concealing them from their enemies, help in thermoregulation and are sometimes, associated with breeding (HAMILTON and BART, 1950).

Females of certain birds have dull colouration. This is due to the greater need for congealment while brooding the eggs. The brilliant colouration of males in many animals does not have any protective value for the male itself, but its conspicuousness might draw attention away from the female on the nest. In some birds, such as Wilson's phalarope, the females are brightly coloured and the darb males perform the job of incubating the eggs. The brilliant breeding plumage of the male is often replaced by the duller darb during the winter season. The light is, thus involved in colouration through its effect on reproduction and through its role in protective resemblance.

Effect on Eyes: The degree of development of eyes sometimes depends upon the intensity of light available. In animals living in caves (*e.g.* Proteus anguinus) and in deep sea fishes, the eyes are absent or rudimentary because these animals live in complete darkness. In surface dwelling forms such as crustaceans and fishes *e.g.* Labeo and Catta the ratio of eyes to head is considered to be normal. In the ocean with increasing depth, size of eyes goes on increasing with the progressive decrease in light intensity.

However, below the upper limits of lightless zone, there is a gradual decrease in the size of the eyes. Some deep sea (benthal) fishes have well developed and enlarged eyes to see in bioluminescent light. Terrestrial nocturnal animals such as owls and geckos have large eyes to see in the dark.

Effect on Vision: Higher animals including man are able to see various, objects only in the presence of light. According to BIGELOW WELSH (1924) and CLARKE (1936) many fishes (*e.g.* Lepomis) depend on sight to locate their food.

Effect on Locomotion: In certain lower animals the locomotion is influenced by light. This is known as photokinesis. For example, the blind larvae of mussel crab, Pinnotheres, move faster when exposed to increased light intensity. Movement of flies is considerably influenced by the wave lengths of light. Locusts stop their flight when the sun is hidden by the clouds.

Phototaxis: In some animals, light plays a role in the orientation of locomotion. This phenomenon of movements of animals in response to light is known as phototaxis. When an animal moves towards the source of light, *e.g.* Rantara and Euglena, it is known as positively phototactic. The animals like earthworms, slugs and certain zooplanktons such as copepods are negatively phototactic as they move away from the source of light.

Phototropism: When only a part of an organism moves in response to light, It is known as phototropism. Phototropism is of common occurrence in plants. Among animals, the hydroids or polyps of many coelenterates and tubicolous worms show phototrophic response.

Photoperiodism: Photoperiodism is the response of an organism to the duration of day-light or length of the day, *i.e.* the time between sunrise and sunset which is known as photoperiod. Between the equator and polar circles, the photoperiod vanes with the season from nearly twenty four hours to nearly no time at all. In temperate regions photoperiod ranges from approximately six to eighteen hours with longer day in the summer and shorter in winter. In equatorial regions the day lasts, for about twelve hours. But it is always the same for a given season and locality. The photoperiod is the most important ecological factor which triggers the physiological; and reproductive behaviour in both plants and animals, such as flowering in certain plants, moulting, fat-deposition, migration and breeding in birds and mammals and the onset of diapause in insects.

Effect of Photoperiod on Plants: In plants a longer photoperiod ensures a greater amount of photosynthesis and subsequent growth. A longer photoperiod leads to an increase in the day temperature and, therefore, increases the metabolic activates of the plants.. Therefore, it could be said that in plants there is a delicate balance between metabolic processes in the light and in the dark. The reproductive activities of most of the plants are also correlated with the length of the day. Based upon photoperiodic responses, plants could be classified into :

(i) *Long day Plants:* These bloom and produce seeds in summer on increasing day-length (*i.e.* more than 12 hours).

(ii) *Short day Plants:* These bloom and form seeds when day light period is less than 12 hours. These plants bloom in spring.

(iii) *Day Neutral Plants:* These are those plants whose flowering is not affected by day-length.

Effect of Photoperiod on Animals: Many animals are also directly affected by changes in the photoperiod. The most obvious is the increased reproductive activity of many birds and mammals; migration of birds, mammals and insects; general condition of plumage; food storing behaviour in squirrels; diapause in insects; and number of nitrogen fixing bacteria in legume plants.

(i) *Increased Reproductive Activities:* The increasing day-length increases gonidial development in fishes, birds and mammals. Most of the animals begin to reproduce during spring period so that their young ones could get optimal thermal conditions. In certain fishes longer days are said to induce early maturity (SUNDARAJ, 1972). ROWAN (1921) and WOLFSON (1953, 60) have shown that certain birds migrate towards the south during winter and towards the north during summer. Recently photoperiodism and its effects on domestic poultry and dairy farming are being exploited by poultry keepers and dairy farmers. If hens are kept in artificially lighted pans during winter in order to increase the length of the day, the egg-laying habit of the hens is stimulated. It is believed that light stimulates the pituitary whose secretion in turn stimulates the gonads.

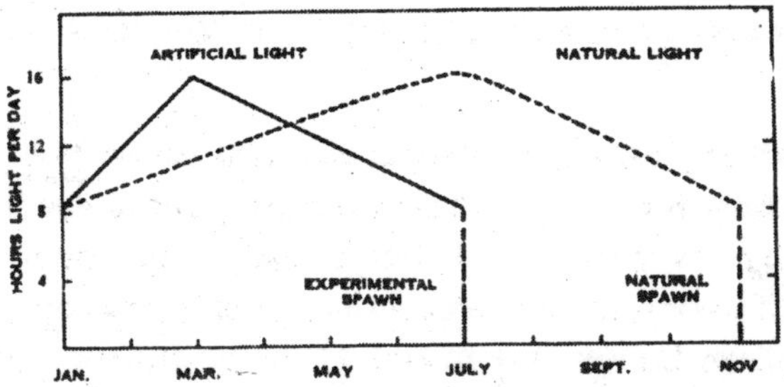

Control of the breeding season of the brook trout by artificial manipulation of the photoperiod (Redrawn from Hazard and Reddy, 1950).

Trout which normally breeds in autumn, spawns in summer when day-length is increased artificially in the spring and then decreased in the summer to autumn conditions.

(ii) *Migration:* The movements of migratory animals such as eels, salmons, birds, mammals and even insects are also thought to be triggered by the photoperiod. Each bird has an internal rhythm adjusted to a definite day-length and a bird migrates to such areas where that particular day-length is available.

(iii) *Diapause:* It was WHEELER (1893) who first coined the term diapause. ANDREWARTHA, (1952), defined diapause as a stage in the development of certain animals during which morphological growth and development is suspended or greatly retarded. During diapause in an animal, growth and reproduction stop, metabolic activities are reduced and resistance to extremes of climate develops to the maximum.

Diapause is of common occurrence in insects during winter. It may occur in any stage of development of an insect. The adaptive value of diapause lies in the fact that it enables the insect to preserve itself for more favourable periods and to resist the extremes of climate.

The larva of pink cotton bollworm found in southern United States goes into diapause in the months of September and October and remain in this stage during winter. Emergence from diapause begins after spring when the days are slightly loner.

But when these larvae were exposed to artificial light in the laboratory, it was found that these larvae could be prevented from going into diapause if this artificial light period ranges from 13-15 hours. In Hieroglyphus nigrorepletus eggs are laid in the soil on bunds and raised grounds during September, October. These eggs remain dormant till the next monsoon in June or July. The hoppers on emerging feed and develop on grass. In the red-hairy caterpillars of groundnut, pupal diapause extends for 9 months in soil.

In grasshopper, diapause occurs in the egg stage. The eggs laid down in soil develop upto the blastoderm stage. Thereafter, no development occur. Any further development occurs only if eggs are exposed to low temperatures. Kharpa beetle exhibits low intensity diapause. In this case, when the temperature drops below 30°C, the larvae do not complete growth and development due to lower metabolic rate. But, at the same time they move about, feed a little and sometimes even moult. The diapause is terminated when the temperature rises from 30 to 35°C.

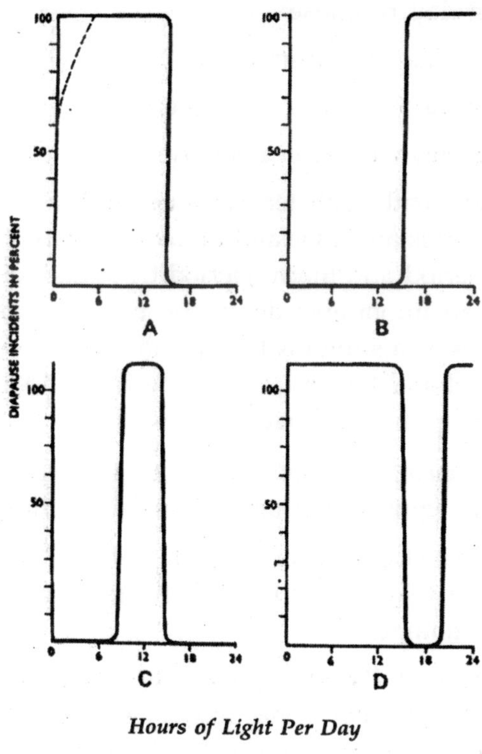

Hours of Light Per Day

Categories of diapause incidence-day length relationship response; among insects.

Photoperiod is an important factor in inducing diapause, since the intensity of diapause is directly related to day-length. In some species the critical day length is within a narrow range of 10-14 hours, while in others the incidence of diapause increase with the increase of day light.

During winter the fall in temperature is associated with a reduction in the number of hours of day light. The diapause is broken by longer day length. This can be illustrated by the diapause of Grupholitha (DICKSON, 1949). Those organisms that entered diapause were found to have hatched during the seasons of the years when there was little more than 12 hours of day light.

A. Long day response;

B. Short day response;

C. Short day-long day response;

D. Long day-short day response.

Long days with high temperatures and short days with low temperatures are associated in the induction of diapause. Thus photoperiod and thermoperiod through their combined action play an important role in the induction of diapause. Food also acts as a stimulus for diapause. Non-availability of a specific food and the variation in the fat and water content may also bring about diapause in insects.

(iv) *Food-storing Behaviour:* The food-storing behaviour of flying squirrel is photoperiodically controlled.

(v) The development of fur and feather coat in animals during winters has been found to be correlated with the photoperiod.

Mechanism of Photoperiodism: It is suggested that the day-length acts through some sensory receptor. The sensory receptor might be eyes in animals and some special pigment in the leaves of plants. The stimulated sensory receptor, in turn, stimulates one or more hormones or enzyme systems that bring about the needed physiological or behaviour response.

Lunar Periodicity: It has been shown that the reproductive cycle of certain different types of organisms show a correlation with the phase of moon. The occurrence of lunar periodicities is well illustrated by the fluctuation in the abundance of conjugants produced by a ciliate living as an ectoparasite on the gills of a freshwater mussel. The distinct peaks occurred regularly on the days following the new moon. These peaks were not correlated with temperature or other environmental changes. Peaks in the population were seen to occur very regularly on the days following the new moon (RAY and CHAKARVARTHY, 1934).

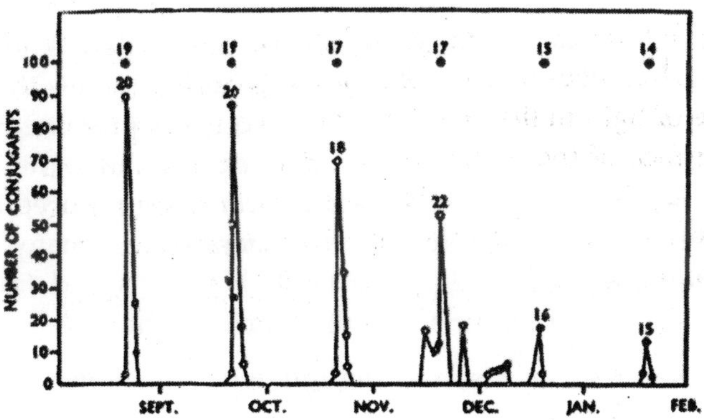

Average daily number of conjugants of the ciliate Conchopnthirius lamellidens on the gills of the freshwater mussel, Laemllidens marginalis. Dates of new moon (top) and of peak numbers are shown.

Most of the organisms exhibiting lunar periodicities are marine. Hence, they are affected by the amplitude of the tide which is greatest at the times of new moon and full moon (spring tides) and smallest at the times of the quarter moon (neap tides). The marine alga, Dictyota, produces its gametes at the time of full-moon spring tide.

The spawning and swarming of a number of marine polychaete worms show a very distinct lunar periodicity coinciding with lunar periods. The palalo worm found in the waters of the South Pacific island comes to the surface in great number on the last quarter of the moon during October and November, producing a luminescence and discharging eggs and sperms into the water.

The worms swim about in small circles in dense masses, giving the sea an appearance of spaghetti soup. In Eunice, swarming occurs during the first and third quarters of the lunar cycle when the intensity of light is low. The Bermuda fire worm puts on a similar display of fire works in the shallow water early in the evening at the time of full moon. It is due to the drop in light intensity following sunset (HUNTSMAN, 1948).

Bioluminescence: Bioluminescence, or popularly known as phosphorescence, is the light of biological origin. It is the only source of light in the deep sea. It is frequently prominent near the surface of the ocean and on land during the night, but it rarely occures in freshwater. Luminescence is produced in sea by certain fishes, crustaceans, coelenterates and many other invertebrates, and on land particularly insects and fireflies. Many groups of bacteria are also luminescent.

Bioluminescence serves the following functions:

1. Illumination
2. Recognition
3. Lure, and
4. Warning.

1. *Illumination:* The luminescence of deep sea animal provides illumination for the individuals producing it as well as for other inhabitants. On land and in shallow water, during the night animals may use their photophores as lanterns.

2. *Recognition:* In the deep sea, fishes locate each other by the pattern of lights presented by their luminescent organs. Squids are able to keep together during the dark hours of the night by means of their characteristic flashing. Luminescence is also employed in the recognition of sex, as exemplified by firefly.
3. *Lure:* Animals lure prey by emitting light from their luminescent organs. The predatory fish attract small fish and planktonic invertebrates towards their luminescent organs which are located near the jaws.
4. *Warning:* The sudden flash of luminescent organs may act as a warning to scare off predators. Certain deep-sea shrimps discharge a luminescent secretion into the water. When the cuttlefish is attacked, it discharges its black secretion into the water and escapes from its enemies in the "smoke screen" so produced.

6

Production of Honey

Man had started the use of animal products since time immemorial even at the cost of animal's life. Honey has been under use in human civilization since prehistoric period as mentioned in our religious literatures like Vedas, Puranas, Ramayana, Mahabharata and Charak Sanghita. Some foreign travellers like Fahiyan and Huen-T-sang had also discussed the use of honey as medicine. People were very much dependent upon honey for medicines and essential nutritive elements of the diet.

The highly evolved social organisation of bees had been established before the existence of human race. Bees teach us the lesson of work-work and work with cooperation. We can easily imagine about the hard-working of the bees by the simple fact that for one pound of honey, a single honey bee travels about double the distance of the circumference of the earth's globe.

Previously, the method of extraction of honey from the honey comb was very much crude but after the invention of artificial hive by Longstroth (1951), it became scientific and commercial. The bee keeping in United States, Canada, Australia and New Zealand has achieved out-standing success.

In India people do not take interest in bee keeping from commercial point of view but only for their routine use. Since honey is produced by the honey bees, a detailed study of the biology of the bees is essential for successful implementation of apiculture programme.

Honey Bee

Phylum .... Arthropoda

Class .... Insecta

Order .... Hymenoptera

Family .... Apidae

Genus .... *Apis*

Habitat and Habit: an Understanding

Honey bees are highly organised social insects reported from all over the world. Although they are active throughout the year but in winter season they do little work and do not rear the brood. In spring seasons *i.e.,* at the time of flowering they prepare a strong colony with honey rich combs. They exhibit polymorphism and good division of labour. The bee hives with thousands of individuals are observed hanging down from the branches of the trees and ceilings of houses. The workers communicate informations for the location of the food sources through the 'Waggle Dance', a phenomenon called as 'Language of the bees', by the eminent biologist Karl Von Frish. He has mentioned that the rate of dance is directly proportional to the distance of the food.

Honey Bees: Species

Four species of honey bees are reported:

Apis Dorsata F. (Rock bee): This the largest bee, about 20 mm. in length, so named as GIANT HONEY BEE. SARANG and BOMBARA are other names of this bee which yields maximum amount of honey in comparison to other species. A single comb may yield 60 pounds of honey which is the maximum amount for a comb. The workers are very smart and active which may pollinate 12,000 flowers daily. But due to its ferocious and irritable nature, specific hive and migratory habit it is very difficult rather practically impossible to domesticate them for the bee keeping industry.

Apis Indica F. (Indian bee): Commonly found in forest and plain regions of India. This is slightly smaller than *A. dorsata.* They prefer to live in dark places and construct several parallel combs about one foot across the protected places like cavities of tree trunks, mud walls, earthen pots, thick bushes, wells and walls of the buildings. This species is very gentle in nature, so can be domesticated easily. The production of honey is much less *i.e.,* 6 to 7 pound per comb.

Apis Florea F. (Little bee): This is smaller than *A. indica* and yields very small amount of honey. The bees are not of gregarious nature and form a single comb. Due to its docile nature and rare stinging behaviour the combs can be removed easily for the honey extraction.

Apis Mellifera F. (European bee): Although this bee produces less honey yet it is found to be the best species from the commercial point of view. Due to their docile nature they can be domesticated easily and can be improved by breeding for several hundred years. Out of its several varieties, the Italian variety is reared everywhere in Europe and America in artificial hives for honey.

Organisation of Society

A highly organised division of labour is found in the colony of honey bees. A good and well developed colony of bees had 40 to 50 thousand individuals consisting of 3 castes *viz.,* QUEEN,

DRONE and WORKER. The queen after fertilization lays fertilized and unfertilized eggs both. From unfertilized eggs male bees emerge which are known as DRONES whereas from the fertilized eggs worker bees are produced. The workers when feed on ROYAL JELLY, develop into QUEEN.

Queen: It is a well developed fertile female provided with immensely developed ovaries. Commonly one queen is found to be present in each hive and feeds on Royal Jelly. She is the queen in real sense as the Mother of the Colony, guarded by a number of attendants and never allotted any duty except egg laying. Egg laying is the sole function of the queen throughout her active life span.

The queen is 15 to 20 mm in length and can be easily distinguished by her long tapering abdomen, short legs and wings. Structurally she is unable to produce wax or honey or gather pollen nector. By the combination of ovipositor-cum sting, a structure is developed which aids in egg laying. It is said that the queen gets mated only once in her life but in a single chance of mating, drone releases 2 crore sperms which are sufficient for the fertilization of the eggs at the time of laying by the female throughout her life span.

In recent researches in U.S.A. it has been reported that out of 110 queens only 55 mated twice before egg laying. It is also a fact that queen lays fertile and unfertile eggs both in accordance to her will but the factors governing such selective activity are still not known. One queen lays about 1,500 to 2,000 eggs in a day depending upon the seasonal variation and other ecological factors.

The total weight of 100 eggs is equal to her body weight. In the whole life span of two to five years a queen lays about 15,00,000 eggs. When the queen in a colony looses its eggs laying capacity, another worker of the same colony starts feeding on queen's diet *i.e.*, Royal Jelly and develops into a new queen and is provided with the facilities of real queen. At the same time old queen may be driven out but sometimes some workers object that as to why the mother of the colony be driven out so ultimately they also come out with the mother. Sometimes when 2 to 3 queens are developed in a colony, only one takes the position of the real queen and the others come out with some workers to establish new colonies.

Workers: Although the workers are the smallest of the three castes but they function as the main spring of the complicated machinery like honey bee colony. Like the queen, they are also produced from the fertile eggs laid by the queen and live in a chamber called as 'WORKER CELL'. It takes 21 days in the development from the egg to the adult and the total life span of a worker is about 6 weeks.

The workers are atrophid female which sacrifice themselves for the well-being of the colony. The total indoor and outdoor duties of the colony are performed by the workers only. That is why they are provided with some special structures for particular work.

(1) Long proboscis for sucking the nectar.

(2) Strong wings for fanning.

(3) Pollen baskets for the collection of pollen.

(4) Powerful sting to defend the colony against any attack.

(5) Wax gland for wax secretion.

The workers which are engaged in outdoor duties, collect the nectar, pollen, gum and water which are received and stored properly by the house bees. The indoor workers are further sub-grouped for specific duties. Some of them which are very sincere, attend the queen while some others look after the nursery called as NURSERY BEE.

Some produce wax for the formation of the new hive and are known as BUILDER. The repairing of the comb is done by the REPAIRERS. The dead body and other impurities are removed from the hive by the CLEANERS. The fanning in the hive is performed by the wings of the FANNERS. Several other functions like honey storage and ripening are also done by the workers. The guard bee always watches at the gateway. It is said that upto half of the life period workers perform indoor duties and later on become engaged in outdoor duties.

Drone: The drone is the male member of the honey bee colony which fertilizes the queen so called as KING of the colony. They take 24 days to develop from the egg to the adult stage. The sting and the wax glands are absent but in the males the reproductive organs are very well developed. They are reared from an unfertile egg in large DRONE CELL. Drone are totally dependent on the workers and have been seen begging for honey from the workers. The sole duty of the drone is to fertilize the virgin queen. At the time of swarming the drone follows the queen, copulates and dies after copulation.

History of Life

After mating the queen generally lays one egg in one brood cell. The eggs are pinkish coloured, elongated with cylindrical body generally attached to the bottom of the cell. Larvae emerge out from both the fertilized as well as unfertilized eggs. Thus, the larvae from the unfertilized eggs form the drones while the workers are developed from the larvae of the fertilized eggs. Amongst the larvae of the workers one is fed on the royal jelly, a special diet secreted by the young workers in the colony, and becomes the queen of the colony.

The royal jelly consists of digested honey and pollen, mixed with a glandular secretion into the mouth of the workers. After 5 days of feeding the cell is sealed and the larvae undergo pupation. It spins a thin silken cocoon and pupates completely. Emergence of the young ones takes place after three weeks and they get busy in the indoor duties for about 2 to 3 weeks. Later on they are sent for the outdoor duties. All the bees pass through a complete metamorphosis with the various changes in the life-cycle taking place within the comb (Fig.).

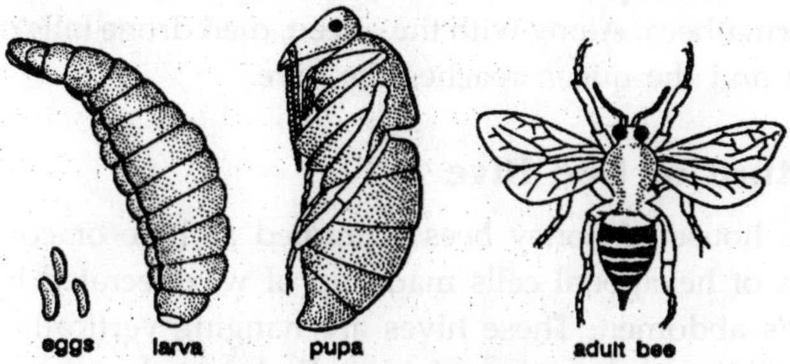

Life history of Apis indica.

Swarming: The process of leaving off the colony by the queen termed as swarming. It happens towards the end of spring or early summer but the real cause of swarming is still not well known. In summers when plenty of food is available and hive is overcrowded by the bees, the queen leaves the hive on a fine fore-noon with some old drones and workers and establishes a new colony at some other place. Now in the old hive a worker is given Royal Jelly and is converted into a new queen of the colony. This new empress of the colony never tolerates her successor, as a natural law in the hive, so she orders to kill the other sisters, if any, in the hive.

Supersedure: When the egg laying capacity of the old queen is lost or it suddenly dies, a new young and vigorous queen takes the position of the old queen and is called as supersedure.

Absconding: The migration of the complete colony from one place to another takes place due to some unfavourable conditions of life, such as destruction of the comb by termites or wax-moths and scarcity of nectar producing flowers around the hive. This phenomenon is quite different from that of swarming.

Nuptial or Marriage Fight: The first swarm is led by the old queen but the second swarm is led by the 7 days old virgin queen which is followed by the drones and is called marriage flight. One of the drones starts copulating with the queen in the sky and fertilizes the queen and dies during the course of copulation. The queen receives spermatophores and stores in the spermatheca. Along with the queen, died drone falls on the ground and the queen reaches the hive.

Structure of Bee Hive

The house of honey bees is termed as hive or comb. It consists of hexagonal cells made up of wax secreted by the worker's abdomen. These hives are hanging vertically from rock, building or branches of trees. Each hive has thousands of hexagonal thin walled fragile cells arranged in two opposite rows on a common base. The resins and gums secreted from the plants are used for the repairing of the hives. The young stages are generally occupying the lower and central cells in the hive which are the 'BROOD CELLS'. In *A. dorsata* broad cells are similar in shape and size but in other species brood cells are of 3 types *viz.*, WORKER CELL for workers, DRONE CELL for drones and QUEEN CELL for the queen.

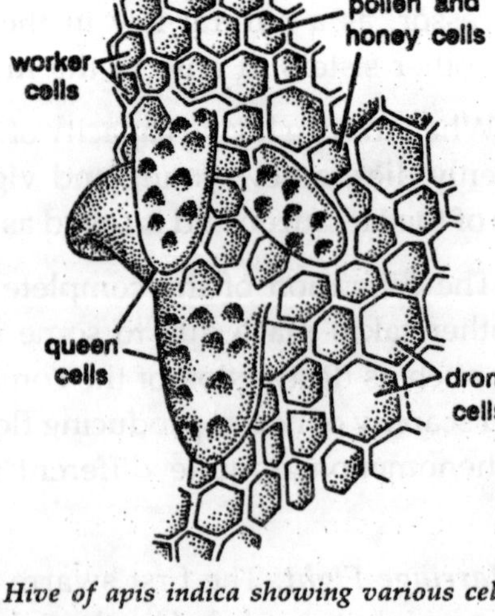

Hive of apis indica showing various cells.

The queen cell can not be used again while the rest are used a number of times. There are no special cells for lodging the adults which generally keep clustering or moving about on the surface of the comb. The cells are mainly intended for the storage of honey and pollen specially in the upper portion of the comb while those in lower part are for brood rearing (Fig.).

Apiculture and Flora

Although honey bees can collect nectar and pollen from quite a long distance but the flora for apiculture is also important. The flora may be of wild or cultivated type. The more nectar yielding plants are neem, jamun, soapnut etc. The other plants like maize, rose and sorghum are good sources of pollen. Some plants like plum, cherry, apple, sheesham, coconut, guava, mustard etc., are good sources for nectar and pollen both.

Bees Selection

For running a good apiary, selection of honey bee is of much importance, so the following should be kept in mind at the time of selecting honey bees for apiculture:

(1) Honey bees should be of gentle temperament.

(2) Honey bees should have capability to construct strong colony.

(3) It should have ability to protect from enemies.

(4) Honey bee should have energetic and industrious workers.

(5) Workers can suck juice from numerous varieties of plants.

(6) Bees on the whole can produce more and more honey from its comb.

(7) Bees can from their comb easily at any place.

The apiculture scientists engaged in genetics are trying to find out such cross races which would not be of ferocious nature but be a good honey producer. In India, *Apis indica* is the best bee for apiculture industries due to its gentle nature and having efficient and prolific workers.

Bee Keeping Techniques

The ultimate aim of bee keeping is to get more and more honey in pure form. The old method commonly used by old apiculturists is very crude, cruel and of unplanned type. This old method is called as Indigenous method.

Indigenous Process

Hive: Two types of hives are used in indigenous method of bee keeping *e.g.*, wall or fixed hive and movable hive.

(a) *Wall or Fixed Hive:* It is purely natural type of comb because the bees themselves prepare the hive at any space on the wall or trees. There is an opening on one side through which bees come out of the hive.

(b) *Movable Hive:* It comprises of hollow wood logs, empty boxes and earthen pots etc. placed in varandas of houses. There exist two holes, one is for entrance and the other for exit of the bees. The swarmed bees usually come to the box on their own accord. Some bee keepers use to take the clusters of the swarms from a tree and keep them in the hive.

Extraction of Honey: For honey extraction, burning fire is brought near the bee hive at the night as a result of which bees are either killed or they escape off. Further the hive full of honey is being removed, cut into pieces and squeezed to get honey. Sometimes smoking is done so that the bees may escape from their hives.

Drawbacks of Indigenous Method: The indigenous method of bee keeping suffers from a number of drawbacks due to which it is not recommended by present day panel. These drawbacks are:

(i) Honey becomes impure because at the time of squeezing, the brood cells, pollen cells, honey cells and larvae are also extracted.

(ii) The colony becomes weak due to killing of the eggs and the larvae at the time of squeezing.

(iii) Formation of new hive by the escaped bees requires extra energy which effects the yield.

(iv) The activities of the bees can be controlled.

(v) The hivation of bees on the same place is only matter of chance.

(vi) The honey robbers, like, rat, ant, wasp and monkeys may affect the hive easily.

(vii) The race improvement programme may not be applied, so no possibility for the selection of the best bee is there.

(viii) The hazards created by climatic factors can not be controlled.

Advanced Techniques

To overcome the drawbacks of indigenous method an advanced method based on scientific facts has been developed. It has opened a new era for the cottage industry in India and has also given an opportunity for lakhs of unemployed persons to keep them busy in this business. From this cottage industry programme the routine agricultural work may not suffer. First of all care was taken to improve the texture of the hives and during this race hive patterns were introduced in India. The Newton model with 7 to 10 frames (21 × 14.5 cm) in the brood chamber with a shallow super (21 × 6.5 cm sized frames) has been most popular in south, east and central India. Longstroth hive containing 10 frames (44.8 × 23 cm) has been used as a standard hive in Himachal Pradesh, Jammu and Kashmir, and Punjab. In Uttar Pradesh another type of hive has been in use which was evolved at Jeolikote apiary and contained 8 frames (30 × 18 cm). After gaining experience from the above mentioned hives, Indian Standard Institute has standardized the hives of small and big sizes accommodating frames 21 × 14.5 cm and 31 × 20.4 cm respectively.

Nowadays a typical type of movable hive is constructed which is capable of expansion or contraction according to the requirement of the place, season and climatic conditions.

Appliances for Modern Method

(1) Typical movable hive.

(2) Queen excluder.

(3) Honey extractor.

(4) Uncapping knife.

(5) Other equipments.

Typical Movable Hive: An artificial movable hive is constructed by wooden box based on bee space theory (Fig.). The size and number of frames are variable from hive to hive according to the need. A small space is enough to permit the entrance and exit of workers and drones but queen once placed in hive never comes outside the hive. The perforation size on zinc sheet is only of 0.375 cm but the thorax of the queen is 0.43 cm to 0.45 cm, so the queen can never pass through this pore. This typical hive consists of 6 parts as given below:

(a) *Stand:* It is the basal part of the hive on which the whole hive is constructed. The stands are adjusted to make slope for the hive. Due to this slope rain water comes down quickly.

(b) *Bottom Board:* It is situated above the stand and forms the proper base for the hive having two gates in the front position. One gate functions as an entrance while the other as exit.

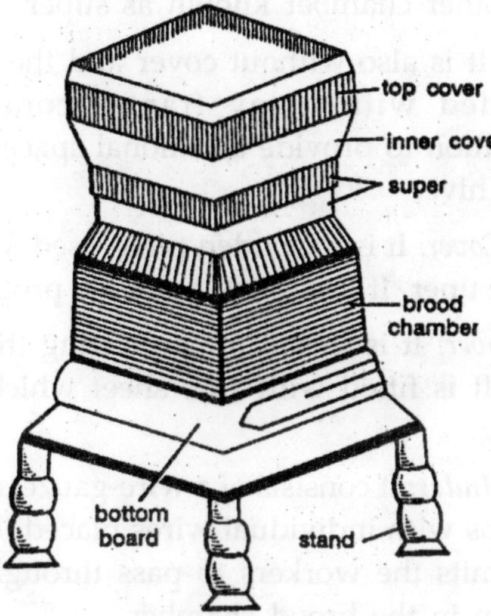

Typical moveable hive.

(c) *Brood Chamber:* The bottom board carries the brood chamber which is the most important part of the bee hive. It is large in size provided with 5 to 10 frames. In each frame a wax sheet bearing hexagonal frames is held up by a couple of wires in a vertical position. Along with the margin of every hexagonal mark, the bees start making wall and ultimately the cells. Here every sheet of the wax is known as COMB FOUNDATION which attracts the bees and provides the base for the comb preparation on both the sides. The frames are kept vertically in brood chamber which is covered over by other frames having a wire meshing through which the workers can easily pass. The comb foundation helps in obtaining a regular strong worker brood cell comb which can be used repeatedly. The

Central Bee Research Station at Pune arranged the manufacture of a comb foundation mill which manufactures, different cell sizes required in several regions of the country. The brood chamber is covered by another chamber known as super.

(d) *Super:* It is also without cover and the base. Super is provided with many frames containing comb foundation to provide additional space for expansion of the hive.

(e) *Inner Cover:* It is a wooden piece used for the covering of the super. It has many holes for proper ventilation.

(f) *Top Cover:* It is meant for protecting the colony from rains. It is fitted with zinc sheet which is plain and sloping.

Queen Excluder: It consists of a wire-gauze, extrans guards and drone traps with individual wires placed 0.375 cm apart. It readily permits the workers to pass through it but keeps back the queen in the brood chamber.

Honey Extractor: It is used for the extraction of the honey from the comb and functions on principle of centrifugal force. When combs are centrifuged by this device the pure honey is thrown out without any damage to the comb.

Uncapping Knife: When all of the combs are filled with honey- they are sealed by capping with the wax. So, before such capped combs are placed in the honey extractor, the wax sealing has to be removed with the help of an uncapping knife heated by steam before use.

Other Equipments: Most of the useful equipments for the successful management of the bee are locally manufactured which are very cheap. As they are made locally, they may not be exactly similar to those made at other places. Thus, Indian Standard Institute has standardized some very common equipments for the production of uniform and interchangeable articles. Some materials like protective garments, gum cages, gloves, net veil, bee net, brush etc. are required for easy and well planned handling of the bees.

Advances of Modern Method: In the modern method of bee keeping there are several advantages which encourage the well planned bee keeping.

(1) A proper watch on the activities of the bees can be had.

(2) A strong colony can be developed by providing sugar, syrup, pollen substances to honey bees.

(3) Swarming of bees is checked by modern hive.

(4) The same hive is used again and again so the workers pay their attention more for the honey and not for the hive formation.

(5) Under adverse climatic conditions the hive can be transferred from one place to the other for the protection of the bees.

(6) Comb can be protected from the enemies.

(7) Pure honey in large quantity can be obtained.

Precautions: For the proper management of bee keeping programme following precautions should be taken:

(1) The hive should not be kept more than half a mile away from the place from where the bees have to collect the nectar and the pollen.

(2) People must know about the bee keeper for proper contact.

(3) The boxes must be kept under shade at cool places.

(4) Industry should be near the road for proper transport facilities.

(5) Fresh water reservoir should be near the hive.

(6) Good flora for the collection of pollen and nectar should be there.

Bee Keeping Products

The chief products of bee keeping industry are:

(i) honey and

(ii) bees wax.

Honey: It is truly an insect product of high nutritive value. The food value of honey may be estimated by the presence of about 80% sugar in it.

Production of Honey: One should not be confused that honey is a direct plant product because the nectar, pollen and cane-sugar bearing secretions of flowers are ingested by honey bees, get mixed with the saliva and undergo certain chemical changes due to enzyme action. At this stage cane-sugar (sucrose) is converted into invert sugars *i.e.,* dextrose and levulose. At this very time some ingredients of bees are also added to the mixture and reduce the water content. The whole mixture is then collected in the honey sac (crop) until the honey reaches the hive. As the honey bee reaches the hive this compound is regurgitated in the hive cell and is known as the honey. Now honey is concentrated by a strong current of air produced by the rapid beating of worker's wings, crawling over the cells.

Honey is very much sweet in taste and white to black in colour with variable smell in accordance with the juices collected from different flowers.

Chemical Composition of Honey: Honey is sugar rich compound having the following constituents:

(1)	Levulose	—	38.9%
(2)	Dextrose	—	21.28%
(3)	Maltose & other sugars	—	8.81%
(4)	Enzymes & pigments	—	2.21%
(5)	Ash	—	1.0%
(6)	Water	—	17.20%

Storage of Honey: After long duration in the stored condition, the honey may be granulated and fermented.

(a) *Granulation of Honey:* The stored honey becomes granular after long duration. Such type of granulation property is the best evidence of pure honey. It is considered that 10 parts of dextrose combine with one part of water, hence forms crystals. Due to less solubility levulose is not crystalised and gives cloudy appearance. Crystallization is mainly accelerated by the presence of minute air bubbles, colloids and pollens.

(b) *Fermentation of Honey:* After crystallization honey is subjected to fermentation. Due to crystallization of dextrose 9% moisture is released, which dilutes the remaining levulose of the honey and the action of yeasts present in air, flowers and soils takes place on levulose and dextrose resulting in the fermentation of honey.

$$C_6H_{12}O_6 \rightarrow 2C_2H_5OH + 2CO_2$$

Economic Importance of Honey: Honey is used by human beings in different ways of which the most important is as food and medicine.

(a) *Food Value:* It is estimated that 200 gm of honey provides as much nourishment as 11.5 litre of milk or 1.6 kg cream or 330 gm meat. 2.1 gm of honey provide as much as 67 K. cal of energy. Its sugars, minerals, vitamins and other vital elements are readily absorbed by the systems. Honey may be taken by healthy men as well as those who are ill. It can be taken at any time in any season and by persons of all ages even those just born. It is used in the preparation of candles, cakes and bread. In illness it is preferred over milk because more than half of the body energy is provided burning of dextrose.

(b) *Medicinal Value:* Honey is mildly laxative, antiseptic and sedative, generally used in Ayurvedic and Unani systems of medicine. It is quite helpful in building up of the haemoglobin of the blood and also used as preventive against cough, cold and fever, as blood purifier and as a curative for ulcers on tongue and alimentary canal. Its regular use is recommended after severe cases of heart attack for malnutrition, indigestion and diabetes. It is also found that typhoid germs are killed by honey within 48 hours, those of branchio-pneumonia in 4 days and of dysentery in 50 hours.

(c) *Other Uses:* Other than food and medicine, honey is used in numerous ways. It is used in the preparation of bread, cake and biscuits. It enhances their preserving quality. Much amount of honey goes in making alcoholic drinks. In poultry and fishing industries honey is widely used. In laboratory, honey is used to stimulate the growth of plants, the bacterial culture, inoculation of seeds of cloves, in insect diet and in the preparation of poison baits for fruit flies.

Importance of Beeswax

Beeswax is a very useful by-product of bee keeping industry. It is yellowish to greyish brown in colour and insoluble in water but completely soluble in ether. Commonly it is a wrong impression to suppose that honey bees convert the pollen into beeswax because beeswax is also a natural secretion of the worker bees and is poured out in thin delicate scales or flakes. Chibnall (1934) has reported that all insect waxes are complex mixture of varying proportions of:

(1) Even numbered alcohols ranging from C_{24} to C_{36}.

(2) Even numbered normal fatty acids from C_{24} to C_{34}, and

(3) Odd numbered normal paraffins ranging from C_{23} to C_{37}.

The various beeswaxes differ only due to change in the proportions of these constituents. Large quantities of beeswax produced and exported, come from *Apis dorsata* bees. Indian Standard Institutions have fixed standards for pure beeswax in order to facilitate its export.

Economic Importance of Beeswax: Beeswax is used in the manufacture of cosmetics, for Catholic churches, face cream, paints, ointments, insulators, plastic works, polishes, carbon paper and many other lubricant. It is also used in the laboratory for microtomy with the common wax for block preparation of tissues.

Enemies of Bees

Enemies of the bees harm the colony in different ways so they have attracted considerable attention in the different regions of the country.

The wax moths *(Galleria mellonella and Achroia grisella)*, Wasp *(Vespa* spp. *and Palarus* sp.), black ants *(Componotus compressus)* and bee eaters *(Merops orientalis)* and Kingcrow *(Dicrurus macrocercus)* are common enemies of the honey bee's comb and honey. Man is the last but worst enemy of honey bees.

Before 1958 bees were considered to be free from the diseases though suspected cases of NOSEMA from Punjab and Kashmir were known. But a parasitic *mite-Acarapis woodi* Rennie caused Acarine disease in the adult honey bee in Kulu valley in Punjab in 1956. It was later reported from Himachal Pradesh, Uttar Pradesh and Jammu and Kashmir.

This disease was controlled by the scheme in cooperation with the United States of America at the college of Agriculture Ludhiana, Punjab. Nowadays Indian honey bees are commonly free from any such disease. A strict quarantine measure is being taken to check the spread of any disease from foreign countries. But in European countries bees are commonly attacked by microsporadian which is injurious to bees.

Trade of Bee-keeping

Bee-keeping has gained a good position as an industry in U.S.A., Canada and Australia but in tropical countries it is not growing as per need of the people. Before 1953 attention to bee-keeping was paid only by the State Governments but in the same year, all India Khadi and Village Industries Commission, started to pay attention to it and it was controlled by Union Government itself.

Due to the functioning of the central organisation, bee-keeping industry was spread in South India and in some northern states also. Nowadays beekeeping industry is nation wide and is a good source of cottage industry. Till now limited research and development work has been done in this field which indicates that 'Apiculture' can be raised to the status of a viable occupation in tropical climates provided appropriate scientific and developmental efforts are generated.

Because the tropical countries are very rich in bee fauna, the scientific development of apiculture may be helpful in agriculture and may contribute for providing nutrition, and employment to rural population and may help in raising the economic status of the people.

Modern Attempts

An international conference on 'Apiculture in Tropical Climates', organised by the 'World Crops' in collaboration with the International Bee Research Association (IBRA) was held at London in 1976. In this Conference it was recommended that the research and development of apiculture should be taken on priority basis in the tropical areas.

The Second International Conference on Apiculture in Tropical climates was organised by the Indian Council of Agricultural Research in Collaboration with the Khadi and Village Industries Commission, Department of Science and Technology and Indian National Science Academy at New Delhi in 1980.

In this Conference Eminent National and International scientists and experts realised the importance of apiculture and suggested further research development as a device for raising the industrial status of apiculture.

7

Production of Silk

For more than 35 centuries countless generations of silkworms are continuously breeding, feeding on mulberry leaves, spinning their cocoons and dying, an everlasting sacrifice to the demand of human-beings for decoration. For the first time in 2697 B.C. Lotzu Empress of Kwang-Ti discovered the fancy origin of beautiful silk in the form of threads.

Thus, the technique of cloth preparation from the cultivated silk was known to Chinese people for more than 2,000 years ago. The art of silk preparation was kept as top secret as national policy and any one, who attempted to send the eggs of silkworm out of the country, was hanged to death. It is said that silk was as valuable as gold and ultimately gold was flowing into China from all parts of the world.

But in the year 555 A.D. this secrecy was opened and the eggs of silkworm and sericulture technique were smuggled by two monks sent as spies to China and thus, sericulture was introduced into Europe. Later on it was introduced in Mediterranean and Asiatic countries. Nowadays sericulture has become one of the most important cottage industries in a number of countries like Japan, China, Rep. of Korea, India, Brazil, Russia, Italy and France.

The experimental and systematic study on sericulture was started in Japan in 1911 after establishment of the Sericulture Experiment Station and in 1979 the National Sericulture Experimental Station was set up in Tokyo. In India first of all Lefroy (1905-1906) started investigation on the silkworm and sericulture at Pusa Institute, New Delhi. He set up a sericulture stall in an 'All India Exhibition' organised by the Government of Uttar Pradesh at Allahabad in 1901-1910 to draw the attention of scientists and general public towards the sericulture industry.

Importance of Silk Moth

The silk producing machine is an insect called as silk moth locally in Hindi 'Resham-ka-kira'. Although a number of species are found producing silk but only few species are used for sericulture industry.

Systematic Position

Phylum ....Arthropoda

Class ....Insecta

Order ....Lepidoptera

Family ....Bombycidae and Saturniidae

Silkworm Species

Mulberry Silkworm: *Bombyx mori.* It belongs to the family Bombycidae. China is the native place of this silkworm but now it has been introduced in all the silk producing countries like, Japan, India, Rep. Korea, Italy, France and Russia. Since the natural food of this worm is mulberry leaf, it is called as mulberry silkworm. The silk produced by this moth is white in colour.

Tasar Silkworm: *Antheraea paphia.* It belongs to the family Saturniidae and common in India, China and Sri Lanka. The caterpillar feeds on ber, oak, sal and fig plants. The cocoon produced by this moth is hard and of hen's egg size which produces relabel brown coloured silk.

Though it had been only a wild variety of silk moth since long, now, by cross breeding, it has been possible to produce such varieties which are reared any how and domesticated. But the domestication of tasar caterpillars is not so easy, so the cocoons have to be collected from the forest. The moths do not easily breed in captivity. Since breeding is not well controlled, the tasar silk industry has not reached upto mark as the mulberry silkworm industry.

Muga Silkworm: *Antheraea assama.* It also belongs to the family Saturniidae and semi-domesticated in nature. The native place of this species is Assam where, it has now become a good source of cottage industry. The caterpillars of this worm feed on *Machilus* plant and the silk produced by this moth is known as Muga silk.

Eri Silkworm: *Attacus ricinii.* It also belongs to the family Saturniidae and produces silk in East-Asia. In India sericulture scientists are trying to produce silk in East-Asia. Sericulture scientists are trying to produce such cross breeds which can provide good quality of silk and can be reared easily. It feeds on castor leaf. Cocoons cannot be reeled as in mulberry cocoons, therefore, it has to be spun. Its life history resemble with that of mulberry worms. The cocoons of this worm have very loose texture and the silk produced is called as 'Arandi' silk' locally. The threads are not glossy but much durable.

Oak Silkworm: *Antheraea pernyi.* It belongs to the family Saturniidae and is found in China and Japan. *A. roylet* of the Himalayas and *A. yamamai* of Japan have been reared for centuries. They produce good quality of silk.

Giant Silkworm: *Attacus altas.* It belongs to the family Saturniidae and is found in India and Malaysia. It is the largest of the living insects reaching upto eleven inches in wing-span.

Out of the above species only four are of common use for sericulture *viz., Bombyx mori, Attacus ricinii, Antheraea assama* and *Antheraea paphia.*

Silkworm of Mulberry

The perusal of literature reveals that much advanced work has been done by sericulturists and scientists on *Bombyx mori* because of its total domesticated nature. Various varieties of silkworm are produced by genetical crossing and they are producing 2-7 generations in India but in some countries like Europe and Russia where the duration of winter is more than the summers only one generation is produced per year.

The race of silkworm by which only one crop is taken in one year is called UNI-VOLTINE, producing two crops in a year is called BI-VOLTINE and producing more than two crops in a year is called MULTI-VOLTINE races. Now- a-days, for commercial purpose only approved races are reared by conducting national trial of silkworm races offered by various research institutions. Different crosses are generally chosen for spring and autumn rearing.

Life History of Mulberry Silkworm: The adult of *Bombyx mori* is about 2.5 cm in length and pale creamy white in colour. Due to heavy body and feeble wings, flight is not possible by the female moth. This moth is unisexual in nature and does not feed during its very short life period of 2-3 days (Fig.)

Fertilization: Fertilization is internal preceded by copulation. Just after emergence, male moth copulates with female for about 2-3 hours and if not separated they may die after few hours of copulating with female.

Egg Laying: Just after copulation, female starts egg laying which is completed in 1-24 hours. One moth lays 400 to 500 eggs depending upon the climatic conditions and the supply of food material to the caterpillar from which the female moth is obtained. The egg laying is always in form of clusters and covered with gelatinous secretion of the female moth which helps them in proper attachment.

Eggs: The eggs laid by the female moth are rounded and white in colour. The weight of the newly laid 2,000 eggs comes to about 1 gm. With the increase in time after laying, eggs become darker and darker day-by-day. Two types of eggs are generally found *viz.*, DIAPAUSE type and NON-DIAPAUSE type. The diapause type of eggs are laid by the silkworm inhabiting in temperate regions, whereas, silkworms belonging to subtropical regions like India lay non-diapause type of eggs. During diapause all the vital activities of the eggs cease.

Hatching: The eggs after ten days of incubation hatch into a larva called as caterpillar. Hatching is the most important phase of silk moth's life. After hatching, caterpillars need continuous supply of food because they are voracious feeders. If proper supply of mulberry leaf is not possible the development of caterpillar would not be in proper course. Sometimes, due to lack of food material, young caterpillars die causing great loss to the sericulture industry. It is recorded that in uni-voltine race hatching of eggs takes one month after laying.

Caterpillar: The newly hatched caterpillar is about 0.3 cm in length and is pale, yellowish-white in colour. The caterpillars are provided with well developed mandibulate type of mouth-parts adapted to feed easily on the mulberry leaves. The caterpillar is twelve segmented and the abdominal region has ten segments having five pairs of pseudo-legs. It is also provided with a small dorsal horn on the anal segment. Because of its being very much tender, the Ist instar larva can feed only on very soft leaves of mulberry plants. As they are voracious feeders, they grow rapidly which is marked by four moultings. After Ist, 2nd, 3rd and 4th moultings caterpillars get changed into 2nd, 3rd, 4th and 5th instars respectively. It takes about

21 to 25 days after hatching. The full grown caterpillar is 7.5 cm in length. It develops salivary glands, stops feeding and undergoes pupation. The time taken for the full growth of the caterpillar from young to the well grown stage varies with regard to the temperature, humidity, food supply and type of race. Weight of the full grown caterpillar varies from 4 to 6 gm.

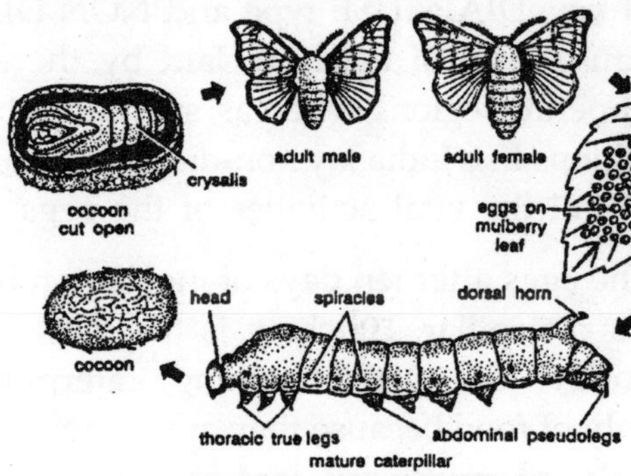

Life history of Bombyx mori.

Pupa: The caterpillars stop feeding and move towards corner among the leaves and secrete a sticky fluid through silk gland. The secreted fluid comes out through spinneret (a narrow pore situated on the hypopharynx) and takes the form of long fine thread of silk which hardens on exposure to the air and wrapped around the body of the caterpillar in the form of a covering called as COCOON.

Cocoon: Cocoon is the white coloured bed of the pupa whose outer threads are irregular while the inner threads are regular. The length of continuous thread secreted by a caterpillar for the formation of cocoon is about 1000-1200 metres which requires 3 days to complete. The threads is wound around the cocoon in concentric manner.

The binding of threads round the cocoon is very interesting and quick going phenomenon achieved by the constant round motion of the head of the caterpillar from one side to the other at the rate of 65 times per minute. Now the silkworm pupa is covered within a thick, oval white or yellow silken cocoon. It is estimated from the data obtained by practical application that one pound of silk can be obtained from 2,500 cocoons. The weight of one cocoon is about 1.8 to 2.2 gm and the weight of the cocoon shell only is 0.45 gm. The size of the thread is 2.0 to 2.8 denier. The pupal period lasts for 10 to 12 days and the pupae cut the cocoon and emerge into adult moths.

Emergence of Imago: Due to active metamorphic changes during pupal period the abdominal pseudo-legs disappear and two pairs of wings develop. The silkworm within the cocoon secretes an alkaline fluid to moisten its one of the ends. As a result of this the moistened end becomes soft where the threads are cut open (hydrolysed) by the silkworm. Finally a hole is formed through which a feeble adult moth squeezes out of the cocoon.

Industry of Sericulture

The production of silk from the silkworm by rearing practices on commercial scale is called sericulture. Although silk is very beautiful and fancy gift of nature but its commercial production is very much complicated and a tough job which requires heavy man power. Before going into the details of the sericulture industry it is essential to have an idea about the major, steps and requirements of sericulture. Improved race of *Bombyx mori* and good nutritive type of mulberry plantation is the sole need for the industry. Other than these some of the other requirements are given below :

(1) *Machana:* The proper place for rearing the silkworms.

(2) *Rearing trays:* For keeping the laid eggs alongwith the mulberry leaves.

(3) *Spinning or Chandrakis tray:* For keeping the caterpillars at the time of full grown stage ready for pupation.

(4) *Dalas:* For fetching the mulberry leaves.

(5) *Chopping knife:* For cutting the mulberry leaves into smaller pieces.

(6) *Baskets:* For the distribution of mulberry leaves.

(7) *Hygrometer:* To know the % humidity in the atmosphere.

(8) *Thermometer:* To take the reading for the room temperature.

(9) *Oven:* To regulate the different stages of life cycle at different temperatures.

(10) *Freez:* For the storage of eggs (seed) for next generation.

Selection of Races of B. Mori: Although a number of races have been developed for the purpose, care should be taken to select only those which would be fit for a particular locality keeping in view the climatic condition of that area. According to rearing season *viz.,* spring, early autumn and late autumn different crosses are chosen for successful rearing.

Mulberry: The whole sericulture industry is based on the best utilization of the mulberry leaves as it is the only food of this insect. In Japan more than thousand varieties of mulberry exist including as many as 130 natural species out of which only a few are in commercial use.

Mulberry is a deciduous medium sized tree belonging to the family Moraceae. Its species are given as below:

Morus Alba Linn: It is of white type and locally called as Toot, Tootri etc. This is widely used for rearing the silkworm and also for fruits.

Morus Indica Linn: It is locally called as 'shahtoot' and is Indian in origin. It is also used for fruits and for rearing of silkworms.

Climate: Although, mulberry is a plant of temperate zone but some varieties have been developed by cross breeding which can grow well in tropical and subtropical zones also. According to the climatic conditions different varieties of mulberry are being used throughout the world.

Soil: For a good crop and nutritive type of leaves, soil should be of loam type and not with stones etc. Soil with neutral pH is best suited for proper growth of mulberry plants.

Plantation: It is of two types as given below :

Seed: In December and May seeds of mulberry are sown and the seedlings grow very quickly which can be transferred either as whole plant or a stump in months of June, July and November. Seed beds should be mixed with ash, lime and white arsenic.

Cutting: Cutting should be done in months of July and August. The size of each cut piece should be 20 to 25 cm in length and 2 to 3 cm in thickness. The cut pieces should be planted in slanting position in the pits (50 × 50 × 50 cm) having 1 kg. superphosphate, 30 gm B.H.C. and 25 kg farm-yard manure.

The plantation is performed in December-January or July. The plantation is mainly of 2 types *viz.*, single row and multiple row plantation. (i) In single row plantation 400 plants can be placed around the boundary of one hectare plot at the distance of one metre from plant to plant. (ii) In multiple row plantation scheme the whole of the plot is used only for sericulture programme. About 10,000 plants are placed at the distance of 1 × 1 metre per hectare. Weekly irrigation is most preferred for bumper growth of mulberry leaves.

Fertilizers: The months of December to January are the best time for manuring the crops. Although the real dose of any fertilizer can be fixed after proper soil testing and according to the fertility of the soil but a generalized recommended dose of manuring is given below (S.S. Sagwal and O.P. Katna, 1981).

Dose of Manures and Fertilizers per Plant

Age of Plant (Year)	*Farmyard manure (Kg.)*	*Kisan Khad (gm)*	*Super phosphate (gm)*
1	10	500	250
2	20	1,000	500
3	30	1,500	750
4	40	2,000	1,000
5	50	2,500	1,250

Pruning and Thinning: December and January are best suited months for pruning and thinning. After 4 to 5 years of plantation mulberry plants get ready for the rearing of silkworms.

The white mulberry *(Morus alba)* or Kashmir white or black mulberry or white Philippine early variety are preferred. In India white Philippine early variety is most preferred because this grows easily.

Silkworm Breeding

The word rearing does not mean only the feeding of caterpillars as often understood but a continuous care from egg laying through aestivation, hibernation, incubation, early stage larval care, late stage larval care to the production of cocoon. So, for proper and step-wise study of rearing one should proceed from Grainage Technology.

Management of Grain

The aim of the establishment of grainage is to provide good quality of seed to rearers and maintenance of original quality of races. For this purpose due care should be taken of the 'crop of silkworm' for seed production from the very beginning *i.e.* the caterpillar stage, by providing them with proper nutrition and protection from the attack of diseases. Keeping these points in view initial selection is made on the basis of percentage of dead pupae during normal development. If it is above the limit seed should not be purchased for seed production. First selection is made by separating out dead cocoons and next selection in the grainage.

After final selection, cocoons are subjected to sex separation by cutting one end of the cocoon either manually or through Nagahara (in Japan) cocoon openers. The Nagahara machine can cut 10,000 to 15,000 cocoons per hour. For the production of commercial eggs loose forms of cocoons are used and collective mother examination is done, either through mass pebrine detecting machine or general microscopic observations. They are then kept for mass emergence.

Emergence of Moth and Fertilization: When kept for emergence at room temperature, mass emergence of adults takes place. As per their nature, just after emergence, male moth starts moving around the female. Males are very much active whereas the females which are loaded with eggs are incapable of flying. If not separated at once in cages males start copulating with the females but the eggs obtained from this female mated from the male of the same stock is useless for the seed.

So the males and the females just after emergence have to be separated into separate cages without their mating. Now one female of one lot is kept with the male of the other lot and atones they form pair and copulate for about 3 hours. After completion of mating, males should be separated and may be used for the fertilization of other females. But one male can not fertilize more than two females. Now fertilized females are subjected to egg laying.

Egg Laying: Just after fertilization, female starts egg laying and in the duration of 24 hours it completes egg laying process. The eggs laid by one female are about 400 to 500 varying according the different races. Female dies after egg laying. These eggs are called as SEED. These eggs are kept in sterilized trays and stored at 4°C under laboratory conditions or sometimes kept at hill stations in diapause conditions.

The stages of egg (seed) production is of 3 types *viz.*, production and supply of grand parent eggs, production and supply of parent eggs, rearing of parent eggs and production of commercial F_1 seed. The grand parent and parent eggs are produced by recognised and reputed organizations. This commercial seed is supplied to the rearers.

Hatching: This is an important phase of sericulture industry because as soon as the larvae are hatched they start feeding voraciously. So only those sericulturists who would be able to supply sufficient amount of fresh mulberry leaves to the young hatched larvae, could perform successful sericulture programme otherwise young ones will die resulting great loss to sericulture industry. This is why the hatching has to be controlled, accelerated or postponed by artificial treatments under refrigerated conditions.

For proper hatching of seeds (eggs) advanced techniques have been developed in which eggs are collected and kept with mulberry leaves, working as stimulant for hatching in shady places on white sheet of paper in insect proof trays on a stool. For this purpose the legs of stool must be kept in water so that insects may not crawl and damage the hatching eggs. It is also notable that if the eggs are placed in the same position in which they are laid, hatching will not be 100 per cent. So it is advisable that the eggs kept in trays should be moved with the help of feather. The group of caterpillars hatched at various stages should be kept separately. Thus, one should be careful that hatching must be coincided with the best season of the mulberry.

Experimental Data of Egg Laying and Hatching: (i) If after 120 minutes of oviposition eggs are kept at 10° C for 24 hours then transferred to the oven at 16°C for 4 days and further soaked in hydrochloric acid (15% at 46° C) for 5 minutes, the diapause condition is broken easily. (ii) To get the homozygous stock of female silk moth, eggs are taken from 24 hour old moth and treated with hot water (46° C) for 18 minutes then kept at 15 to 17° C for 4 days, treated with acid at 46° C and then kept for incubation. When such eggs are reared carefully a homozygous stock is obtained. (iii) If the female moth is kept at 5°C for 24 hours just after mating and for egg laying at normal room temperature, 70 per cent eggs are laid in the first hour after incubation and 20 per cent eggs are laid in 2nd hour. This data is very much useful for grainage technicians from the point of view of uniform hatching of eggs.

Rearing as a Trade

After grainage management the next step is the supply of seed or caterpillars to the fanners. The supply is of two types depending on the knowledge of rearers *i.e.*, supply of eggs and 2nd instar larvae.

The old rearers who are well versed with the rearing technique may purchase eggs for the rearing but new and untrained rearers knowing nothing about the rearing should always be given 2nd instar caterpillars for this purpose. Much care should be taken for the rearing of 1st, 2nd and 3rd instar caterpillars and 4th and 5th instar caterpillars are mostly reared either on hanging trays often with nylon nets or on the floor.

It is essential to clean the bed of the caterpillars once in a day for 3rd, 4th and 5th instars. The principle of rearing should be the production of healthy caterpillars with uniform development. The mulberry leaves supplied to the 1st and 2nd instar larvae should be young and well chopped because they are very tender in nature. The 3rd instar larvae may feed even on chopped hard leaves but 4th and 5th in star larvae can feed easily on leaves also.

The taste of mulberry leaf is lost due to rapid loss of moisture, so rearer should be careful to stop or reduce -the loss of moisture from the leaf after harvesting. For this purpose 60 to 70% humidity of the room is needed otherwise it affects the health of silkworm.

So, it is advisable to keep the leaves wrapped in wet clothes. The overcrowding of worms causes under-nourishment and checks proper development. If a large number of rearers in the same village take the worms from same lot, spinning of silkworm of all the rearers starts on the same day if proper supply of mulberry leaves is there.

Now full grown 5th instar larva stops feeding, undergoes pupation and settles to. a corner among the mulberry leaves and starts to secrete sticky fluid. Thus, an improved technique of rearing of silkworm has resulted in the production of cocoons of good quality. The temperatures best suited for the rearing of 1st, 2nd, 3rd, 4th and 5th instar larvae are 27, 27, 25, 24 and 23° C respectively.

Cocoons Spinning

This is the period when the caterpillar stops feeding and starts to secrete a pasty substance from the silk gland. In this condition worms should be picked up and transferred to the spinning trays and kept in a position of slope (slanting) to the sun for a short period. Within three days spinning is over and the cocoon is formed and this is the last phase of the rearing of silkworm.

Quality of Cocoon: The quality of cocoon is dependent on the raw silk yield, filament length, reliability and splitting.

Marketing of Cocoon: The price of cocoon is fixed during every season of the rearing. This price is, however, watched by the Government and cocoons are purchased by the rearers.

Processing after Cocoon

The method of obtaining silk thread from cocoon is known as post-cocoon processing. This includes STIFLING and REELING.

Stifling: The process of killing the cocoons is termed as stifling. Sericulturists should be very much careful that before the emergence of silkworm (The cocoons which do not have cut holes) good sized cocoons of 8 to 10 days old are selected for further processing and dropped into hot water or subjected to steam or dry heat, sun exposure for 3 days or fumigation. In this way pupae or cocoons are killed. The killing of the cocoon in boiling water helps in softening the adhesion of the silk threads among themselves and loosening of the outer threads to separate freely, facilitating the unbinding of silk threads.

Reeling and Spinning: The process of removing the threads from the killed cocoon is called as reeling. Four or five free ends of the threads of these cocoons are passed through eyelets and guides to twist into one thread and wound round a large wheel from which it is transferred to spools. Thus the silk obtained on the spool is called as RAW SILK or REELED SILK. The waste outer layer or damaged cocoons and threads are separated, teased and then the filaments are spun. This spunned silk is called as 'SPUN SILK'.

The raw silk is further boiled, stretched and purified by acid or by fermentation and then carefully washed over again and again to bring about the well-known lustre on the thread.

The modernization of the reeling and spinning process by autolization and various labour saving process has opened a new way to this cottage industry in the world. One autolizer can yield 3.715 kg silk per basin in 8 hours in Japan.

Silk: Silk is a pasty secretion of the silkworm produced by the silk gland. The silk glands are actually modified salivary glands which are long and sac like. As this pasty secretion comes in contact with air, it becomes hard and forms strong and pliable silk strands. This secretion forms two cores of fibroin: (i) a tough elastic insoluble protein consisting of 75% of the fiber's weight and cemented together with sericin from the middle region of the silk gland at the time of secretion, and (ii) a gelatinous protein which is easily soluble in warm water. Some quantity of wax and carotenoid pigments are also detected. The diameter of the silk fibers is 0.0045 to 0.0082 cm. Its elasticity is found to be 20%.

Silkworm Diseases

The Sericulture Industry suffers from a number of diseases in Tropical regions of South East Asia. The maggot disease, pebrine, polyhedrosis and flacherie are the diseases which cause severe damage to this industry. The poisoning by tobacco and Muscardines is also reported to be harmful but is not very common.

Effect of Maggot Disease

This disease is caused by *Tricholyga sorbillans,* a fly belonging to the order-Diptera and Family-Tachlnidae. It is distributed throughout India, Japan, Korea, China, Vietnam and Thailand. The presence of milky white cylindrical eggs on the skin of silkworm larvae is a symptom of this disease. The number of eggs laid by the fly on a silkworm larva varies from several to more than fifty but usually they are two to three. After 30-40 hours of egg laying, maggot is formed inside the egg shell. Now this maggot makes a hole on the ventral side of egg shell and on the skin of the silkworm.

Thus, maggot penetrates into the body of silkworm larvae and starts eating the tissues of larvae. When the maggot penetrates into the larval body, the big black mole is formed on that part of the skin. The segments in which maggot exists, swell up and bend as a result the attacked larva becomes inactive and loses appetite. Usually fourth and fifth stadia are attacked by the maggots. The larvae, attacked upto the fourth stadia die before making cocoon, whereas, those attacked in the fifth stadium make cocoons but usually do not attain the pupal stage.

A number of natural enemies like parasitic insects, fungi etc. are known to control the population growth of *T. sorbillans.* The avoidance of entry of this fly into the rearing room of silkworm is the only means for the prevention of this disease for which nets are fixed around the windows.

What is Pebrine?

This is one of the worst diseases of silkworm. Sericulture was once damaged by this disease in all the countries involved in this industry but some countries have overcome this disease and succeeded in getting pebrine-free-silkworm-eggs for reeling cocoons. *Nosema bombycis* Nageli, belonging to Microsporida, is the casual micro-organism of this disease.

The infection takes place through the mouth of larvae at the time of feeding or through the mother's ovary. When new spores are formed in the tissues of alimentary canal of silkworm, they are discharged with the faeces and make a source of infection. When the spore enters into the digestive tract of silkworm, two nuclei, contained in the sporoplasm, are divided into four nuclei and at the same time the polar filament projects and penetrates into the cells of the alimentary canal.

Further, they enter into the blood and swim in it. They are distributed throughout the host body, attacking various tissues, specially the fat bodies and organs, excluding chitinous tissues and nucleus of cells. When the hypoderm is attacked and its cells die, the affected part becomes black due to the formation of melanin.

Inside the body, the milky white spots or marks are observed on the silk gland or on the surface of the alimentary canal. In case of severe infection of the eggs the whole of yolk nearly gets filled with the micro-organisms resulting in their (eggs) death. Whereas, in case of slight infection the eggs hatch but the larvae carrying infection die at the third moult without making cocoon.

The diseased larvae show little inclination towards food and exhibit irregular and differential growth resulting in the formation of small larvae of different sizes. They become tardy, shrunken or moult quite late and finally die.

Counter measures against pebrine. If the infection occurs at the embryo stage, the larvae die in the third moulting stage as already mentioned. So, during the rearing of reeling cocoons it is very important that the pebrine free eggs are used.

The infection by pebrine should be strictly avoided at any stage of the life history of the silkworms for rearing seed cocoons in order to get the healthy eggs. Therefore, it is essential for the prevention of pebrine that the tools used and the rearing houses should be kept free from pebrine germs. The following measures may be applied.

Examination of Mother Moths: All the mother moths producing eggs should be carefully examined one by one, and eggs laid by healthy moths only be taken for further use. The unqualified eggs should be burnt away.

Forecasting and Correcting Examination: In order to make the pebrine examination more reliable, the forecasting and correcting examinations are carried out for the eggs of the seed cocoons. The materials employed for forecasting examinations are the excrement of mature larvae, late moulting larvae, dead larvae, cocoons or pupae and moths accelerated to emerge out of cocoons. For the correcting examination a few eggs of each reproductive egg-batch are taken and incubated and thus, emerged larvae are used as the material for examination of pebrine germs.

Removal and Disinfection of Pebrine Germs: The spores of pebrine may survive for a number of years in an ordinary type of rearing room if the environmental condition is humid. Therefore, rearing rooms, tools and other utensils should previously be cleaned and washed to remove the pebrine diseased eggs, carcases of infected larvae, pupae, moths and dead cocoons, faeces of diseased larvae and so on. The pebrine spores can be destroyed by treatment with 2% formalin for 30 minutes, 0.5% sublimate for 5 minutes, 5% chlorinated lime for 30 minutes, current steam for 30 minutes and sun shine in the summer for 7 hours.

Care on Rearing Silkworms: The tendency of the infestation by the pebrine spores has been observed to occur more when silkworms are reared in dry and cool conditions than in the hot and wet conditions. If the larval period becomes longer, the infection of pebrine is severe, so care should be taken on these points during the course of selection of the seed cocoons by the rearers.

Silkworms and Polyhedrosis

Three types of polyhedrosis are found in silkworms:

(1) Nuclear polyhedrosis.

(2) Intestinal cytoplasmic polyhedrosis.

(3) Intestinal nuclear polyhedrosis.

1. *Nuclear Polyhedrosis (Grasserie, Jaundice):* It is caused by a kind of virus which forms polyhedra in the nuclei of the cells of fatty tissues, dermal tissues, muscles, tracheal membranes, basement membrane, epithelial cells of midgut and blood corpuscles. The polyhedra are commonly hexagonal and rarely tetragonal in shape, containing large number of virus in them. The

polyhedra form the white pus after they are released into the body fluid. The virus present inside the polyhedra maintains its pathogenic power for a number of years in the rearing room but the isolated virus loses its pathogenic power in a period of short time.

The larvae infected by this virus become inactive and lose appetite, and the membrane between segments swell up. Further, the whole body swell up showing the loose skin. In the last phase of the disease the body becomes purulent and the skin becomes tender from which pus leaks out. The larvae in this phase crawl around up and down and finally die. The infection of grasserie occurs through the mouth of the larvae, the wounds of skin and induction under extremely adverse conditions (cold treatment, heat treatment, chemical treatment).

Countermeasures

(1) The rearing rooms, tools and utensils should be washed, cleaned and disinfected.

(2) The highly nutritive mulberry leaves should be given sufficiently to the infant larvae. The fresh air should be circulated in the rearing room especially in fifth stadium.

(3) The high temperature, low temperature and high moisture content should be avoided especially during the infant stages to maintain good health.

(4) The diseased larvae or dead bodies should be removed and the room disinfected completely by 2% formalin.

(5) The silkworms should be reared carefully, so that they may not suffer from the wound of skin.

Intestinal Cytoplasmic Polyhedrosis: This type of polyhedra are formed in the cytoplasm of midgut cells but in a few cases they are formed in the goblet, too. The polyhedra of this disease contains plenty of virus. The infected cells of midgut rupture and polyhedra are released into the gut. Thus the faeces, excreted becomes whitish, containing plenty of polyhedra.

The symptoms of this disease are the occurrence of translucent cephalothorax and shrinkage of body size but in the advanced phase the larvae excrete white coloured loose faeces. The body fluid of the diseased larvae remains in the normal condition.

The mode of infection of this disease and countermeasures are similar to those of the nuclear polyhedrosis.

Intestinal Nuclear Polyhedrosis: In this disease the virus makes polyhedra inside the cytoplasm and nucleus of the midgut cylindrical cells. The polyhedra formed in this disease are large sized. The translucent cephalothorax, shrinkage of body and diarrhoea are the symptoms of this disease also. The mode of infection and countermeasures are similar to those of nuclear polyhedrosis.

Importance of Flacherie

Flacherie is the genetic name of some kinds of silkworm diseases, carcases of which rot due to the attack of bacteria. Flacherie may be divided as follows:

Infectious Flacherie caused by a Kind of Virus: The various symptoms of flacherie are loss of appetite, translucent cephalothorax, vomiting and diarrhoea but the real diagnosis can be made after the microscopic observation of the virus.

The pathogen of this disease is a spherical virus which does not form polyhedra in the body of silkworm larvae. The virus multiplies in the tissues of midgut and is released into the gastric juice and is excreted alongwith faeces which is the source of infection. The virus infects the larvae of silkworm orally.

Countermeasures

(1) Resistant silkworm races should be selected.

(2) The rearing room, tools and utensils should be well disinfected.

(3) In order to maintain the good health of silkworm, high quality mulberry leaves should be provided.

(4) Favourable conditions like temperature and humidity should be maintained in rearing rooms.

(5) The faeces, diseased larvae and dead bodies should be piled in the compost.

(6) Some of the chemicals like hydrochloric acid, formalin, chlorinated lime may be used as disinfectants for this virus.

Gastric injury caused by physiological disturbance of silkworms followed by the multiplication of bacteria. Due to the supply of bad quality of mulberry leaves the digestive physiology of the silkworm is disturbed and multiplication of bacteria in the gastric cavity takes place.

Thus, the combined action of physiological disturbances and bacterial activity in the gut are major causes of this disease. In unfavourable climatic conditions, the bacteria like *Streptococci* sp. *Coli aerogenous bacilli* or proteus group bacilli attack the weakened silkworms. The control measures against this disease is to keep healthy conditions of rearing silkworms.

Bacterial Intoxication: This disease is caused by a toxin of some bacilli, *Bacillus thuringiensis* Var. The larvae attacked by this toxin become unconscious, later soften, become darkish and finally rot off.

The infection occurs orally and can retain the toxicity as long as for seven years in some cases. The countermeasures are the disinfection of the rearing room and instruments.

Septicaemia: This disease is caused by infection of some bacteria as *Bacillus megatherium, B. proteus, B. prodigious, B. pyocyones* in the blood of silkworms. This disease is rare in the larval stage but it causes severe damage to the pupae and the moths during the period of egg production. The infection is caused through the wounds on the skin.

Countermeasures

(1) The rearing tools and rearing houses should be kept sterilized.

(2) The diseased cases should be kept away and put into the fermenting compost.

(3) During the period of egg laying the temperature and humidity of the room should be maintained in proper order specially at lower range of humidity.

Uses of Green Muscardine

It is a fungal disease of silkworms. There are a number of muscardine in silkworm but only green muscardine *(Spicaria prasina)* has been noticed to affect the larvae of silkworm in Vietnam.

The infection may be observed at the third and fourth stadia of silkworm. In the beginning stage a big black spot is observed on the ventral side.

The green-tubes of fungus develop into mycelia in the blood and bear cylindrical spores which are separated from mycelia and further form mycelia which bear cylindrical spores again. Thus, all the organs of silkworm are attacked by this fungus disturbing their normal functioning. As a result the diseased larvae do not moult and finally die.

Countermeasures

(1) The rearing room, tools and utensils should be disinfected.

(2) The diseased larvae and their faeces should be piled into compost to kill the germs.

(3) The rearing bed should be kept dry as much as possible to prevent the germination of conidia of fungus.

Significance of Economy

The raw silk is used in the manufacture of woven materials and the knitted fabrics for the preparation of garments, parachutes, parachute cords, fishing lines, sieve for flour mills, insulation coil for telephones and wireless receivers, and tyres of racing cars. Fabrics for garments in various weaves, plain, twill, stain, crepe, georgette and velvet, knitted goods such as vests, gloves, socks, stockings, dyed and printed ornamented fabrics for saries, jackets, shawls and wrappers are made out of this material.

Trade in Sericulture

At the root of the social, economic, cultural and political progress of India, there are 6.5 lacs villages where 75% of the population of the country lives. The real progress of our country is definitely on the development of these villages. For the economic independence of the villages it is most important to check the flow of ingenious people from the villages towards cities so that the villages also may get a chance for advancement and progress. Even today the main source of earning livelihood, in villages, is agriculture and agriculture based industry on which 70% of the population depends.

Majority of the village population even today lives below poverty line. It is believed that small scale and cottage industries are only in small and under developed countries but this assumption is only partially correct because in developed countries like USA, Germany, France, Britain, Russia, Switzerland and Denmark cottage industries occupy a special importance in their economy. In such countries where there is derth of capital but ample labour power, small scale and cottage industries also have great importance. When we discuss the economic policy of India, one doubt always haunts our mind, weather on the strength of small and cottage industries only, would it be possible for us to face the challenges in the international economic competition. Gandhi's economic philosophy provides solution to this doubt that small cottage industries and modern and large industrial units have to be free from mutual competition, then only, all round development of the nation can be possible.

In the back ground of the industrial growth of past few years and their popularity and distribution of cottage industries it is apparent that the silk industry has developed as a popular cottage industry in which all big, middle, marginal and small farmers in their own way in their limited resources can start silkworm rearing. Even the landless labours with the cooperation of farmers can get mulberry leaves and rear silkworms. This industry neither needs any large equipment, big capital nor specialized technical know how for its start. Although sericulture is practised in India since long even much before independence, yet because of more capital involved at that time and low quality production it could not achieve required success.

In any region for achieving success in sericulture industry there are some basic factors which should be surveyed. It is most important to have additional knowledge of the climate, soil, previously developed small industry, if any, and economic and social structure of that region. In case the soil and the climate of the region is suitable for rearing of silkworms, proper arrangement for irrigation, availability of labour etc. have also to be thought of. In short it must be fully ascertained weather this new industry of silkworm rearing would be definitely profitable in contrast to the traditional farming or other small and cottage industries, if any, prevalent in that region before embarking on this new project.

The idea of the capital investment involved and the average income per acre should also be calculated so that the farmer be told that it would be profitable to them. If finally decided to undertake this venture then for selecting the race of silkworm for rearing, the climate, soil and irrigation facilities should again be kept in mind so that the rearers may not face any technical difficulty later on.

In the country due to the efforts of sericulture research and investigation cum training centres and other production programmes, there has been a good improvement in the production and quality both of the silk cloth during the last 25 years. In a number of regions new varieties of mulberry with close spacing and proper instructions have proved to be much fruitful in some regions but a number of rearers are still unaware of the proper cultivation technique and use of fertilizers for the mulberry plants. By efforts of systematic researches it has been possible to improve the nutritive quality and growth of mulberry leaves from 15,000 to 35,000 kg per hectare.

The new mulberry varieties K_2 and M_5 are giving 65% more yield than the local crop. Now some strains are capable of giving 100 per cent increase in the yield of mulberry leaves. Sericulture research stations and grainages have evolved techniques for handling of young leaves easily with respect to temperature, humidity, quality, quantity and frequency of feeding, spacing, aeration and protection against diseases. A number of techniques have developed which has resulted in the improvement of the maintenance of silkworms.

Although India occupies second position after China in silk production yet the production in India is one fourth of the total production in China. The increasing demand of silk cloth in India and its decreasing production in Japan has placed China as the biggest market of silk in the world. The ever growing interest shown by Japan in the manufacture of electronic goods has left Japan much behind in silk production. Central Silk Board was established in 1949 in India and since then, through development and research plans, silk industry got a great encouragement.

But in 1989, with the cooperation of 'World Bank', 'National Silk Project' has doubled in silk production in seven years and in the next five years the production may be still doubled from the present quantity. 85% of the silk cloth manufactured in India is utilized in the country itself. India imports raw silk for its industry from China. Out of the total silk cloth woven here, 65% is woven by handloom, 30% by powerloom and only 5% in mills by modern machine.

In India the export quality of silk can only be produced by using threads manufactured by modern machines only. At present 3.5 lacs hectare of land is under mulberry cultivation in India. In Karnataka, Andhra Pradesh and Tamil Nadu the interest of the farmers in silk production is continuously increasing day by day. In the country to day silkworm rearing programme is providing whole time and part time employment to 75 lacs of people.

Growth of Silk Industry

Of the 6.5 lac villages in India sericulture is practised in about 75,000 villages. The production of mulberry silk is confined to the five traditional states like Andhra Pradesh, Karnataka, Tamil Nadu, West Bengal and Jammu and Kashmir. Now it has also been initiated in other twelve states like, Kerala, Maharashtra, Bihar, Madhya Pradesh, Orissa, Gujarat, Rajasthan, Punjab, Harayana, Uttar Pradesh, Assam and Himachal Pradesh and gaining success in some other states.

The traditional states produce about 99 per cent of the total mulberry raw silk in the country. The major production of tasar raw silk is achieved from Bihar, Orissa, Madhya Pradesh, Andhra Pradesh and West Bengal, while Maharashtra and

Uttar Pradesh are producing tasar raw silk on a small scale. The eri raw silk production is being achieved from major eri silk producing states like Assam, Bihar, Meghalaya and Manipur while it is also produced in Arunachal, Mizoram, Tripura, Nagaland and Orissa at small scale.

Manipur, Nagaland, Mizoram and Arunachal Pradesh are the oak tasar silk producing states. The production of muga silk is mostly confined to the states of Assam but nowadays this silk is also produced in Mizoram, Meghalaya and Nagaland.

Project of National Sericulture

For the rapid development of mulberry sericulture industry in India, a five year programme. 'National Sericulture Project' of Rs. 555 crore was launched in 1989 by the World Bank and the Swiss Development Corporation under the supervision of Central Silk Board. This project was implemented by the department of sericulture in five traditional mulberry silk producing states like, Andhra Pradesh, Karnataka, Tamil Nadu, West Bengal and Jammu and Kashmir. Twelve pilot projects in the 12 states like, Assam, Bihar, Gujarat, Kerala, Harayana, Himachal Pradesh, Madhya Pradesh, Uttar Pradesh, Punjab, Maharashtra, Rajasthan and Orissa were implemented by the Central Silk Board to create additional facilities for the silkworm seed production, training, maintenance of germplasm and extension programmes as well as research and Development facilities.

The broad objective of NSP was the introduction of mulberry sericulture in non-mulberry silk states, and expansion of this industry in mulberry silk producing states. The other important objectives of this project were generating employment opportunities for an additional one million people, increasing raw silk production, improving quality of silk produced, strengthening the infrastructure for research, extension, seed production, silk processing and development of market support for raw silk and cocoons. In addition to the above objectives the financial support to rearers, reelers and twisters, increased participation of private sectors, involvement of Non-Governmental Organisations and improved participation of women were also the priorities of NSP.

The specification of National Sericulture Project in Traditional and Pilot states was of different nature on the policy as well as application level. In traditional states the main thrust was towards consolidating the status of progress already achieved by strengthening the existing infrastructure and by providing additional facilities for further expansion. The development of water management technique, involvement of rainfed technology for mulberry, promotion of smokeless Chulhas in reeling units, establishment of grainage, basic seed farms, quality control measures, women welfare and sound Research and Development facilities were the main features of NSP.

The Central Silk Board has strengthened the already existing research facilities at various centres like Central Sericulture Research and Training Institute (CSR and TI) Berhampore, Central Sericultural Research and Training Institute (CSR and TI) Mysore, Central Silk Technological Research Institute (CSTRI) Bangalore and a number of Regional Sericulture Research Station (RSRS). Central Silk Board has also established three new research institutes like Seribiotechnology Laboratory (SBL), Silkworm Seed Technological Laboratory (SSTL) and Silkworm and Mulberry Germplasm Station (SMGS) during the tenure of NSP in traditional states.

Before the launching of the NSP in 1989 mulberry sericulture was already growing in small pockets of some non-traditional NSP states like Uttar Pradesh, Bihar, Assam, Orissa and the North- East but no integrated developmental effort were taking place in absence of infrastructural facilities. The pilot projects of CSB under NSP in pilot states provided comprehensive packages right from basic seed production to the yarn production. The CSB has invested Rs. 660 million in non traditional states from 1989 to 1994.

The pilot activities were restricted to one district (in some cases more) in each non-traditional state to ensure the achievement of the target. The district taken under pilot project were Akola and Buldana (Maharashtra); Ambala (Harayana), Bastar (Madhya Pradesh), Dehradun (Uttranchal), Saharanpur (Uttar Pradesh); Hoshiarpur (Punjab), Jorhat and Sibsagar (Assam); Koraput (Orissa), Palakkad and Idukki (Kerala); Purnea, Kishanganj and Araria (Bihar); Solan (Himachal Pradesh); Surat and Valsad (Gujarat) and Udaipur and Banswara (Rajasthan).

Sericulture Industry and Women Welfare: It is a reality that spite of all efforts the status of women in the industrial sector is not significant because there are so many constraints for a women to participate freely in the industrial activities. Sericulture is such an agro-based cottage industry in which there is involvement of interdependent rural, semi-urban and urban based activities in which the estimated participation of women is about 60 per cent.

Thus, in contrast to other agro-based profession the role of women in sericulture industry is dominating-which will be helpful for improving the status of women in family enterprises.

In the light of women welfare through sericulture industry the Central Silk Board has established a special component of 'Assistance to Women and NGCKs' into the National Sericulture Project. For the women welfare sericulture should lead to formulate positive additions to rural women's awareness, participation in decision making and proportional control over resources and income.

For achieving improved status and recognition of their value and importance to household, community and society as a whole, group formation should be encouraged among the women involved in sericulture industry.

Silk Board at the Centre

Central Silk Board, a statutory organisation under the administrative control of the Ministry of Textiles, Government of India, was constituted in 1949 under an act of the Parliament. After its constitution, Central Silk Board has taken overall responsibility for the intensive development of growing sericulture industry in India. The Board is constituted by 36 members including the Chairman, Vice-Chairman, Member Secretary, representative of both the Houses of Parliament, Central and State Government's nominees, Rearers, Reelers, the Trade and Industry. The head quarter of CSB is in Bangalore.

Function of Central Silk Board

(1) Proper development of sericulture industry in the country through promoting necessary measures.

(2) To promote and encourage scientific, technological and economic research for further advancement of sericulture industry.

(3) To develop healthy silkworm seed and ensure their proper distribution.

(4) To devise advanced means for the improvement of mulberry cultivation, silkworm rearing, reeling of silk and spinning.

(5) To search and establish the means of standardisation and quality control of silk and silk products.

(6) To stabilise the prices of silk cocoon and raw silk, and rationalise the marketing.

(7) To collect the statistics regarding sericulture industry.

(8) To prepare and furnish the relevant reports on sericulture to the Central Government.

(9) To give proposal and advise to the Central Government for the intensive development of sericulture industry in the country.

Coordination of Central Silk Board with States: Being a state subject in the concurrent list, the schemes for the proper development of sericulture industry are formulated and implemented by State Governments. For performing this work, additional money is allocated to the State Governments on the basis of the annual plans approved by the planning Commission.

Besides coordinating the sericulture developmental activities in states and advising the Central Government in policy making programmes, the CSB is also directly concerned with the sericulture research activities, sericulture Post Graduate Training programmes, basic seed production, distribution of raw silk, inspection of silk goods and standardisation and quality control.

The Central Silk Board has established 5 'Regional Offices' and 7 'Regional Development Offices' in different parts of the country. The Regional offices are Bangalore (Karnataka), Mumbai (Maharashtra), Calcutta (West Bengal), New Delhi (Delhi) and Srinagar (Jammu and Kashmir). The Regional Development offices are Bhubaneswar (Orissa), Guwahati (Assam), Hyderabad (Andhra Pradesh), Lucknow (Uttar Pradesh), Chennai (Tamil Nadu), Maheshmati (West Bengal) and Patna (Bihar). Special attention for promoting sericultural programmes in North-eastern region has been paid by CSB and an office of the Director (NE) is functioning at Dispur in Assam.

Silk Development and Research: Though the sericulture is one of the ancient industries in India yet it remained inadequate from technological know how and productivity. The establishment of Central Silk Board in 1949 has been proved to be the backbone for developing sericulture industry in India which is providing the much needed research and developmental support to this growing industry. Within the span of five decades the achievement of CSB is encouraging particularly in the field of research and training programmes. To meet this requirement 'Central Sericultural Research and Training Institute was established in Mysore in 1962. Further, the status of Central Sericultural Research Station, Berhampore, established in 1943, was upgraded into a research institute.

In 1972 the Regional Tasar Research Station on Temperate Oak Tasar came into existence. The state research stations were taken over by the CSB and two Regional Research Stations were established at Titabar and Majra.

The establishment of International Centre for Training and Research in Tropical Sericulture (ICT RETS) in Mysore initiated international cooperation in sericulture development. For conducting research on silk reeling, spinning, weaving and processing of silk the first institute of its kind in the world, Central Silk Technological Research Institute,' was set up in Bangalore in 1983.

There is a wide network of extension centres carrying extensive work for the transfer of sericulture technology from lab to land through out the country. Nowadays a number of Research Institutes and Regional Research stations are actively engaged in sericulture research activities. The list of these institutes and stations is as under:

Silk-Research Institutes

(1) Central Sericultural Research and Training Institute (CSR and TI) Mysore (Karnataka).

(2) Central Sericultural Research and Training Institute (CSR and TI) Berhampore (West Bengal).

(3) Central Tasar Research and Training Institute (CTR and TI) Ranchi (Jharkhand).

(4) Central Silk Technological Research Institute (CSTRI) Bangalore (Karnataka).

Regional Silk-Research Stations: A number of Research Centres have been established in different parts of the country for carrying out research in mulberry, oak tasar, muga and eri silk:

Mulberry Silk-Research Station

(1) Bangalore (Karnataka)

(2) Chamaraj Nagar (Karnataka)

(3) Salem (Tamil Nadu)

(4) Coonoor (Tamil Nadu)

(5) Kalimpong (West Bengal)

(6) Dhule (Maharashtra)

(7) Pampore (Kashmir)

(8) Anantpur (Andhra Pradesh)

(9) Jorhat (Assam)

(10) Ranchi (Bihar)

(11) Koraput (Orissa)

(12) Mothabari (West Bengal)

(13) Majra (Uttar Pradesh)

Oak Tasar Silk Research Station

(1) Batote (Jammu and Kashmir)

(2) Bhimtal (Uttaranchal)

(3) Imphal (Manipur)

Muga Silk Research Station

(1) Boko (Assam)

Eri Silk Research Station

(1) Mendipathar (Meghalaya).

8

Scene in India

The environ scenario of India is very wide indeed. At the first level, special attention must be paid to the school going children and women (about 50% of the population). They are to be made aware of health, nutrition, sanitation, hygiene, development, water and food contamination, fodder and fuel wood etc. Non-government organizations (NGO's) have to pay a significant role. In the Directory of the Dept. of Environment, there are 200 NGO's which work in the area of EE and awareness. Moreover, children are to be told the real meaning of wildlife. They are to be educated for plants, smaller animals, microbes etc. *i.e.* holistic approach to wildlife.

Ecological Awareness

The chief goals of EE in India must be: 1. To improve the quality of environment. 2. To create an awareness among people on environmental protection. 3. To develop the capabilities of decision making. The spectrum of EE has four major interrelated components *i.e.* (i) awareness, that include making the individual conscious about the physical, social and aesthetic aspects of environment, (ii) real-life situations, that link environment to life, these conditions are locational specific, thus problems and properties of each area may be different,

(iii) conservation and (iv) sustainable development, where spot light would be on utilisation and not on exploitation. In the former, the idea is that all resources are finite and there is also a limit to the growth of living systems.

Thus resources are to be used in wise manner. Sustainable development aims at utilization of resources not only by the present but also by the future generations in a manner that utilization (and not exploitation) is balanced. Utilization of resources for development is always associated with some negative impacts. Thus efforts are to be made to contain or minimise them.

Primary School Stage: The attempt is made to sensitize the child about environs. Emphasis should be mostly (75%) on building up awareness, followed by real life situation (20%) and conservation (5%). The contents to be used are surrounding from home to school to outdoor situation. Teaching strategy includes audio-visual and field visits.

Lower Secondary Stage: At this level objective must be real life experience, awareness and problem identification. The quantum of awareness must decrease with increase in real life situations. The contents are supplemented with general science. Teaching, practicals and field visits are to be made.

Higher Secondary School Stage: The emphasis must be on conservation, assimilation of knowledge, problem identification and action skills. Contents may be science based and action oriented work. There should be proper teaching, practicals and field work.

College Stage: Maximum emphasis should be on knowledge regarding sustainable development based on experience with conservation. The contents must be college based on Science and Technology. Teaching practicals and action-oriented field work is to be done. In the school education NCERT has been playing vital role in designing syllabi, text books, guide books, charts, kits, teaching materials and other aids.

University Education: EE at this level is being looked after the UGC. There are about 10 universities teaching environmental sciences. The University education has three major components- Teaching, Research and Extension. At post graduate level, four major areas are recognised:

1. *Environment Engineering:* It includes subjects like architecture, civil engineering, town and country planning, including human settlement, slum improvement, landscape architecture, industrial design, regional science and urban ecosystem studies.
2. *Conservation and Management:* It includes fields like land use, forestry, agriculture, energy, waste management, wild life management, national parks, biosphere reserves, biological diversities, water management, mining management, non-polluting renewable energy development etc.
3. *Environmental Health:* This deals with public health and hygiene, sanitary and chemical engineering, occupational health, toxicology, nutrition and drug use etc.
4. *Social Ecology:* It includes subjects like ecology, sociology, social planning, cost-benefit, community organisation and services, psychology and counselling, environmental ethics and related areas of humanities.

There are some institutes, centres assisted by Dept. of Environment which provide formal education/training in environmental areas *e.g.*, Centre for Environmental Education (CEE), Ahmedabad, Indian Institute of Forest Management, Bhopal and Indira Gandhi National Forest Academy, Dehradun. This education is designed for any age group, participating in cultural, social, economic development of the country. They form clubs and arrange exhibitions, public lectures, meetings, environmental campaigns. Following are the main constitutes of this education.

Adult Education: Adults may influence the society to protect the precious environs by generating posters, slides, audio-visual and information pictures.

Rural Youth and Non-student Youth: They may act as volunteers.

Tribals and Forest Dwellers: They are an important media to protect the forest wealth.

Children Activities: Dept. of Environ with the help of United School Organisations of India organised essay competitions among different age group children. School term courses are also given by NMNH in EE every year. The National Museum of Natural History (NMNH) conducts spot painting, modelling and poster design about environment for children.

Eco-development Camps: A set of guidelines has been prepared by D.O. En (1984). The objectives are: (i) To create awareness in youth about basic ecological principles. (ii) To identify root cause of ecological problems as related to human activities. (iii) To promote for solving ecological problems. (iv) To develop a spirit of national integration.

Non-government Organisations: There are more than 200 NGOs of which most are involved in EE and awareness. Others in pollution control, nature protection and conversation, rural development, waste utilization, wild-life conservation, floristic and funal studies, afforestation and social forestry and eco-development.

Public Representative: India has environmental forums for MPs and MLAs to discuss environmental problems facing the country. They may be sound public opinion and stimulate public interest for saving the environs.

Training Executives: Regular courses should be arranged for environ activities among administrators.

Research and Development Programme: Such R&D efforts are supported by D.O. Environment in Biosphere and

Man as well as solving basic and applied environmental problems.

Foundation Courses: The courses for the probationers selected for IAS, IFS, IPS and cadets of three wings of Armed Forces need to be supplemented with foundation courses on environment relevant to their area of specialisation.

Development of Educational Material and Teaching Aids: Materials for media (T.V., radio films, newspaper etc.), audio, mobile exhibitions, audio-visual materials must be operated by competent manpower. One such centre in India is centre for Environmental Education, Ahmedabad.

Development of Trained Manpower: Dept. of Environment (DOE) must organise training programmes for the professors, technical personnel, lecturers and legal experts.

National Environment Awareness Campaign or National Environment Month: Commencing from 1986, DOEn conducts NEAC and NEM, from November 19th to December 18th every year is observed as NEM. Each year there is major environmental theme.

World Environmental Day: All Govt. in the states, UTs, Universities, Schools, Colleges, academic institutions and voluntary organisations organise suitable activities on WED, *i.e.* 5th June of each year DOE supports the function financially.

National Environment Awareness Campaign/National, Environment Month: Commencing from 1986, Department of Environment conducts NEAC and NEM; from November 19th to December 18th every year is observed as NEM. Each year there is a major environmental theme.

Study Material: The environmental education may be thought not as an end in itself but as a means to an end. Education is the process of development of child's personality. The Review Committee (1977) on curriculum of EE has emphasized the need for stressing more environment based education. The committee has also recommended that physical and bio-sciences, life sciences curriculum should be made environmental oriented with emphasis on problem related to environment.

The review of literature on the theme of environment reveals that it is related to physical, biological and social sciences, thus, the edifice of environment, knowledge and understanding is interdisciplinary in nature. Therefore, it can be adequately designed and included in the course curriculum of teacher education programme. The teaching methodology of physical sciences, bio-science and social sciences are taught and practised under the main course of teacher-education. Therefore, the related concept and components of environment can be easily included in these teaching subjects of teacher education curriculum. The universities and educational institutions have designed the course content of Environmental education for B.Ed, and M.Ed, level. The main emphasis has been given on physical and biological component of the environment.

Tiresome Courses

1. *Historical Development of E.E.:* Ancient view point and recent concept of E.E.
2. *Concept of Environment:* Meaning structure and processes. Types of environment and components.
3. *Concept of Education:* Meaning, objectives, content and status, interdisciplinary approach to education.
4. *Concept of Environmental Education:* Relationship between Environment and Education, meaning and definition characteristics and objectives, content, method and techniques.
5. *Concepts related to E.E.:* Ecology, Ecosystem, Autocology, Biomes and Pollution.
6. *Physical Environment and Education:* Meaning components, Air, Water, Sound/Noise Pollution.

Meaning, structure, components, factors affecting environmental, industry, urbanisation, traffic , etc. and role of education and interdependency of various factors of environment.

7. *Social and Cultural Environment and Education:* Meaning and structure, Pollution, colonisation, urbanisation due to social changes and mobility. Development of media, factors influencing the environment and role of teachers and schools, method health and community health.
8. Agencies of Environmental Protection and Information-Steps taken. The governmental organisation, private organisation, legal steps, role of educational institutions.

Teaching Techniques: As already mentioned that the contents of EE is largely interdisciplinary in nature. It is both art (doing) and science (understanding), organised from primary to university level, the objectives of EE are not confined upto knowledge and awareness but include skills, attitudes and values. Thus strategies of teaching and learning have wide coverage, based on content, its components, levels of education and the objective of E.E.

The major objectives of E.E. are awareness, attitude and action. Therefore, a student should be allowed and asked to observe simple phenomena of earth and sky. The teacher should also translate awareness in his action which will be the model for the students. Thus the doing part much more important for the teacher. The role of teacher is very significant in realizing the above objectives of E.E. A number of projects can be assigned to the students in school situations for making herbarium, plantation in school campus. It is the teacher who can sensitize his students for improving the quality of environment. From the given table it is evident that E.E. employs scientific and non-scientific methods of teaching. All the methods provide the awareness but skills and attitudes are equally important objective in teaching environment education.

Strategies of Teaching of Environmental Education with Reference to Objectives and Levels.

Strategies of Teaching	*Objective of Environmental Education*			
	Awareness	*Skills*	*Attitudes*	*Action*
Primary Level				
1. Observation method	+ + + +	——	++——	——
2. Playway method	+ + + +	+——	+ +——	
3. Field trips	+ + + +	+——	++——	——
4. Dramatisation	+ + + +	——	++——	+——
Secondary Level				
1. Lecture method	+ + + +	——	++——	——
2. Questions-Answer	+ + + +	——	——	——
3. Project method	+ + ——	——	——	——
4. Educational tours	+ + + +	——	++——	——
5. Dramatization	+ + + +	——	++——	++——
6. Observation	+ + + +	——	+++——	——
Higher Level				
1. Lecture method	+ + + +	——	++——	——
2. Group discussion	+ + + +	——	++++	——
3. Seminar and w.s.	+ + + +	+++——	+++——	++——
4. Survey method	+ + + +	——	+++——	——
5. Action research	+ + + +	+++——	+++——	++++——
6. Experimentation	+ + + +	++——	+++——	++——
7. Interdisciplinary approach.	+ + + +	——	++++——	++——
8. Demonstration	+ + + +	++——	++++——	++——

The objectives of E.E. can be realized through formal and non-formal system of education. The strategies of E.E. have been summarized in the above table:

Practical Application: Ryan developed 'Educational Excursion Method' of teaching which is commonly used in geography, botany, zoology, study of nature and history teaching.

There are a number of topics in E.E. which can be taught effectively by this method.

Role of Technology

The following are the theoretical basis of this method.

(i) It provides the real experience about the E.E. to the learners.

(ii) It involves the observations imagination and ability of discovering the cause effect relationship among the environment components.

(iii) It develops the feeling of cooperation and group work as social principles.

Objectives

(i) To develop the tendency of excursion and its utility for the understanding.

(ii) To provide awareness about the environment.

(iii) To develop the ability of observation, imagination and discovering.

(iv) To develop ability of cooperation and team work.

The following steps are taken is organising educational excursion.

(i) The specific objectives are to be formulated.

(ii) A schedule—date, time, number of students, duration and financial aspects should be finalized.

(iii) A guide sheet is prepared and supplied to the students.

(iv) The students are asked to prepare a report about their observations and visiting different places.

(v) The visiting organisations are to be contacted for the arrangement.

(vi) The excursion should be followed group discussion to highlight the environmental components and source of pollution.

Uses

(i) Teaching can be made effective and interesting by real experiences.

(ii) It helps in developing ability of observation and imagination.

(iii) The objectives of teaching can be realized by this method.

(iv) It develops feeling of cooperation and team work.

The co-curricular activities of education are the most appropriate means for providing opportunities for the action. The teacher is the main means for implementing the programmes and realizing and objectives of E.E. and organisation of co-curricular activities in schools and outside the school.

There are two main programmes recommended by Educational Commission and policy *i.e.* National Social Scheme (NSS) and socially useful productive work (SUPW). The activities organised through these programmes are:

(i) To clean the environment through NSS camps.

(ii) To grow plants and develop gardens for protection.

(iii) To clean the public places and parks etc.

(iv) To construct roads and dig pits for the wastes.

(v) To develop sense of sanitation among the people by organising cultural programmes.

(vi) To develop the consciousness about population education by organising camps of population education or family planning.

(vii) To encourage the students to prepare charts related to environmental pollution and its protection.

(viii) To motivate the students for using stories and essays on the environmental education.

The Review Committee (1977) on curriculum E.E. has stressed the need for more environment based education. There are more than two hundred non-government organisations engaged in environmental education awareness and training youths on different areas.

Recently the university departments are organising seminars and workshops on E.E. to develop the course content for B.Ed and M.Ed classes.

The courses are to be designed at two levels - first in the core course and second is the specialized course. The core course is compulsory for B.Ed students and specialized course is the part of teaching methodology course.

Various Programme

1. Aswan High Dam in Egypt (1964). It is the biggest dam in the world. It was constructed at Aswan in 1964 on river nile. It has caused several environmental problems of serious consequences because the likely or probable impact of dam and reservoir on environmental conditions were not properly assessed. It caused several kinds of biotic and abiotic problems.

2. Earthscan. An agency founded by UNEP in 1976 that commission's original articles on environmental matters and sells them as features to newspapers and magazines especially in developing countries.
3. Convention on International Trade in Endangered Species. (CITIES). An international forum whose membership for agreement is open to all countries. For India, the Ministry of Environment and Forests functions as nodal agency for participation in international agreements.
4. Environmental Protection Agency (EPA). This is an independent Federal Agency of the U.S. Government established in 1970. It deals with protection of environment by air, water, solid wastes, radiations, pesticides, noise etc.
5. European Economic Community (EEC). It is a community of 12 European nations with sound political, economic and legal base. The community has joint agricultural and scientific programme between its members. It also assists other countries in environmental programmes. It has programmes of framing and implementation of coordinated policy for environmental improvement and conservation of natural resources. CPCB, India has recently taken up a project on air quality monitoring with assistance of EEC.
6. Human Exposure Assessment Location (HEAL). The project is a part of the Health Related Monitoring Programme by WHO is cooperation with UNEP. The project has three components *viz.* (i) air monitoring (ii) water quality monitoring and (iii) food contamination monitoring on a global basis.
7. International Council of Scientific Unions (ICSU). A non-governmental organisation, based in Paris, that

encourages the exchange of scientific informations, initiates programmes requiring international scientific cooperation and studies and reports on matters to social and political responsibilities in treatment of scientific community.

8. International Union for Conservation of Nature and Natural Resources (IUCN). An autonomous body founded in 1948, with its Head quarters at Morges, Switzerland, that initiates and promotes scientifically based conservation measures. It also cooperates with United Nations and other inter-governmental agencies and with sister bodies of world wildlife fund (WWF).
9. Trans-Alska pipe line in USA (1977). This project was initiated to carry mineral oil and natural gas through 1270 kms long underground pipelines from the newly discovered oil wells near prudhoe bay on north Coast of Alaska to port Valdez to south coast of Alaska. This project has not affected Alaska and marine and environment except the political problems that the route and pipe line involves Canada. It is not advisable due to safety and strategic point of view.
10. International Marine Consultative Organisation (IMCO). It regulates the operation of ships in high seas, from marine water pollution viewpoint.
11. United Nations Educational, Scientific and Cultural Organisation (UNESCO). A United Nations agency founded in 1945 to support and implement the efforts of member states to promote education, scientific research and information and the arts to develop the cultural aspects of world relations. It also holds conferences and seminars, promotes research and exchange of informations and provides technical

support. Its Headquarters are in Paris. Independently as well as in collaboration with other agencies like UNEP, it supports activities related to environmental quality, human settlements, training to environmental engineers and other socio-cultural programmes related to environment.

12. United Nations Environment Programme (UNEP). An UN agency, co ordination of inter government measures for environmental monitoring and protection. It was set up in 1972. There is a Voluntary United Nations Environment fund to finance environmental projects. There is an Environmental Coordination Board to coordinate the UNEP programmes. Its Headquarters are in Nairobi, Kenya. UNEP was founded to study and formulate International guidelines for management of the environment. UNEP is assisting many such programmes in India.

13. Earth watch Programme. A worldwide programme, established in 1972 under the terms of the declaration on the Human Environment. It monitors trends in the environment based on a series of monitoring stations. Its activities are coordinated by UNEP.

14. Man and the Biosphere Programme (MAB). The programme is the outcome of International Biological Programme (IBP) that has already concluded its activities. MAB was formally launched by UNESCO in 1971. There are 14 project areas under this programme. We shall provide here the details of MAB with special reference to major activities in our own country so far done under the same, and the priority areas for future.

15. Project Earth. Developed in collaboration with UNEP to inspire, interest and educate young people world wide on crucial issues facing the Earth's Environment.

The project is led by Mr. Robert Swan, UNEP Good will Ambassador for youth. He is the only person to have reached the North Pole and the South Pole on foot.

16. Tehri High Dam Project (India). It is one of the most controversial river projects in world as regards the conflict between environmentalists, local people, project authorities and government. It is being constructed on the Ganga river below the confluence of its two main tributaries-Bhagirathi and Bhilangane at Tehri of Uttar Pradesh. Tehri Hydro corporation (THDC) was formed in 1989 and construction work was taken over by Union Government of India with soviet technical and economic aid. It is 260.5 meter highest rock fill dam in the country. Before planning this dam, proper assessment was not done to visualize the probable adverse effects.
17. Sardar Sarovar (SS). Project near Navagam in Bharauch district of Gujarat is one of the costliest projects affecting villages in three states - M.P., Maharashtra and Gujarat. About 245 villages will be submerged of which about 193 in M.P. alone. Over 75000 people will be evicted. Additional displacement is likely to be caused during social and environmental rehabilitation work undertaken to repair the dislocation and damages caused by the project.
18. Narmada Valley Project (NVP). The world's largest river valley project has attracted the greatest attention. The 30 big dams and over 3000 medium and minor dams are envisaged at cost of Rs. 25,000 crores. It would displace one million people, mostly tribals, submerge 56,000 ha of fertile agriculture land. Total forest area of nearly 60,000 ha. will be destroyed . About 25 species of birds will be deprived of their habitats.

19. Bodhghat Project. On Indravati river in M.P. is in heavily forested Bastar district. The project will destroy teak and sal forests. The criticism of the project forced the Government and the World Bank to reconsider it.

Man under Effect: MAB in fact is the outcome of the experience of those involved in the International Biological Programme (IBP). It was realised that several problems require collaboration of natural and social scientists, planners and managers and the local people. MAB was conceived at the International Biosphere Conference at its 16th session in 1970.

The programme was formally launched by UNESCO in November 1971, when the MAB International Coordinating Council held its first session and identified 13 project areas of cooperative research. One more area of project area was added in 1974.

Role of Education

Department of Environment could set a plan programme-Environmental Information System (ENVIS) in 1982. It is decentralised system using distributed network of data bases for collection of environmental information.

ENVIS Network with department of environment, its focal point consists of 10 ENVIS centres on diverse area of environment, established in specialised and reputed institutions in the country.

Objectives of MAB: The general objective of the programme is to develop the basis within the natural and social sciences for the rational use and conservation of the resources of the biosphere and for the improvement of the global relationships between man and the environment; to predict the consequences of today's actions on tomorrow's world and thereby to increase man's ability to manage efficiently the natural resources of the biosphere.

Approach: The programme aims to provide scientific basis to solve real practical problems of resources management through understanding the environmental problems in the ecosystem context.

Organisation: Internationally MAB is guided by International Coordinating Council (ICC) but in India, the MAB is serviced and funded by the Department of Environment, Forest and Wild Life. Indian National Man and Biosphere Committee, was first constituted in 1972 which supervise and directs the programme.

Project Areas: Research under MAB Programme is divided into 14 broad themes which are given.

Project Area 1. Ecological Effects of Increasing Human Activities on Tropical and Subtropical Forest Ecosystems

The broad themes include:

(i) baseline studies in natural tropical forest ecosystem.

(ii) problems involved in the management and natural regeneration of exploited tropical forest ecosystem.

(iii) investigations into the methods of regeneration of exploited tropical forest ecosystem.

(iv) effects of various methods of cultivation and land use on soil structure.

Project Area 2. Ecological Effects of Different Land Uses and Management Practices on Temperate and Mediterranean Forest Landscapes

(i) to identify and assess changes due to man's activities on such landscapes and the effects of these changes on man.

(ii) a comparative study of interrelationship between natural and man made forest ecosystems and impact of changes in human population.

(iii) to develop ways of measuring quantitative and qualitative changes in forest environment.

(iv) ecological impact of forest fire and effect of air pollution on forest ecosystem.

Project Area 3. Impact of Human Activities and Land Use Practices on Grazing Lands

(i) to secure information from natural and social sciences research on grazing lands in order to provide guidelines for management of these lands under different climatic and socio-economic conditions.

Project Area 4. Impact of Human Activities on the Dynamics of Arid and Semi-arid Zone Ecosystems with Particular Attention to Effects of Irrigation

The main aim is:

(i) ecological improvement of arid and semi-arid regions with special emphasis on the effects of irrigation.

Project Area 5. Impact of Human Activities on Mountain Ecosystems

The chief objectives are;

(i) effect of shifting agriculture, over grazing, deforestation

(ii) impact of large scale technology such as hydroelectric power and water storage dams

(iii) effect of tourism and recreation

Project Area 6. Conservation of Natural Areas and the Gene Material They contain.

The main area of interest are:

(i) preparation of list of rare and threatened plant and animal species.

(ii) protection in the form of biosphere reserves and National Parks etc.

(iii) development of techniques for rapid multiplication of threatened and rare plant and animal species.

Project Area 7. Ecological Assessment of Pest Management and Fertiliser use on Terrestrial and Aquatic Ecosystem

The programme lays emphasis on development of methods for control of pests of agriculture and public health importance.

Project Area 8. Ecological Aspects of Urban Systems with Particular Emphasis on Energy Utilisation

The area of research are, energy and material flows into the urban areas and pattern of their utilisation.

Project Area 9. Perception of Environmental Quality

(i) perception of environmental hazards.

(ii) perception of National Parks and other relatively unmodified natural areas.

(iii) perception of quality in urban environment.

Project Area 10. Research on Environmental Pollution and its Effects on the Biosphere

The major areas in the project are:

(i) to identify the environmental indicator.

(ii) acute and long term effects of pollutants from domestic, industrial, agriculture or other sources.

(iii) to monitor the level of different pollutants in the environment.

Project Area 11. Ecological Effects of Human Activities on the Value and Resources of Lakes, Marshes, Rivers, Deltas, Estuaries and Coastal zone 5

The main objectives are:

(i) origin, nature and extent of pollution and its effect on aquatic life.

(ii) sewage problems.

(iii) harnessing the river for agriculture, animal husbandry, drinking and industrial water supply.

(iv) tidal movement effect on ecosystem.

Project Area 12. Ecology and Rational Use of Island Ecosystem.

Project Area 13. Effects of Major Engineering Works on Man and his Environment.

Project Area 14. Interaction between Environment Transformation and the Adaptive, Demographic and Genetic Structure of Human Population.

The above project has following objectives:

(i) Folklore survey and collection and identification of plants and animals used by tribals.

(ii) Conservation of plants used by them.

(iii) Impact of tribal culture on vegetation and wild life.

(iv) Technology of development of tribal communities.

A number of research projects on these aspects have been sponsored. A unit of Genetic Toxicology is established at Kolkata University to study (i) the chronic effects of sub-toxic doses of pollutant on human systems in relation with different modifying factors as genetic polymorphism, nutrition, etc. and (ii) identify the species of vegetation genetically tolerant to various pollutants.

A project is initiated to study the effect of SO, and particulate matter on vegetation in the urban and industrial areas. The organisations involved are, B.H.U., Varanasi, J.N.U., New Delhi, N.B.R.I., Lucknow, B.S.I., Kolkata National Organisations.

A number of Government and non-government organisations are engaged in environmental studies. Department of Environment Forest and Wild life of India was set up in 1980 to serve as the focal point in the administrative structure of the Government for planning, promotion and coordination of environmental programmes.

9

International Factors

Environment is defined as "the sum total of all conditions and influences that affected the development and life of organisms." The interplay of material cycles and energy flow in natural ecosystem generates a self-correcting homoeostasis with no outside control or set points. The ecosystem is capable of self-maintenance. However, the equilibrium is very sensitive to external stimuli such as human activities promoted by socio-economic goals.

The global community recognised the seriousness of the U.N. declared 1990s as the International Decade for "Natural Disaster Reduction." As a part of this world wide programme of education and action has called upon to make people aware of the danger ahead and the steps they must take to check and mitigate their ill effects. As we know that Global climate is changing due to green house effect" *i.e.* due to the accumulation of green house gases such as carbon dioxide, chloro-fluorocarbons (CFCs) methane, nitrous oxide and ground level ozone gas.

Half of the global warming is alone by carbondioxide gas. Temporary stability can be achieved if annual emission of CO, could be reduced by shifting fossil fuel use from coal to oil to natural gas. This can also be achieved if we shift to other alternate sources of energy like solar, geothermal, tidal, hydro etc.

The ozone layer present in stratosphere protects us from the ultraviolet rays which would cause different ailments like sunburns, skin cancer, eye damage etc. This is also depleting because of released of CFCs, nitrous oxide and other gases. In order to avert the worst effect of warming due to green house effect serious efforts are being made all over the world to reduce the release of these gases into the atmosphere.

Preservation Issues

Conservation has been defined as "the management for the benefit of all life including human kind of the biosphere so that it may yield sustainable benefit to the present generation while maintaining its potential to meet the needs and aspiration of the future generation."

As the earth is the only planet where life can exist, population on the earth is increasing at an alarming rate which is putting stress on the natural resources at much larger scale. Continuous increase in population caused an increasing demand for resources. This created a situation when the non-renewable resources may come to an end after some time.

Basically there are two main reasons for exploitation of environment and its resources. One reason is poverty while the other is rich people. Although development is necessary for man but it should be in a way that it should not destroy the resources.

All the available resources available on the earth should be used judiciously so that they should not be depleted. The people should be made aware of the consequences of this depletion. Since the natural resources are over exploiting there should be a check or control over it.

Aims and Purpose

The main objectives to conserve the natural resources are:

(i) to maintain the proper ecological balance and life support system.

(ii) to preserve biological diversity.

(iii) to ensure that any utilisation of species and ecosystems is sustainable.

Conservation, therefore, makes important contributions to social and economic development.

Making of Plans: At global level there are only two strategies

(a) Special Interest Conservation Strategies. Conservation is to be done of only important and limited resources.

(b) Total Ecosystem Conservation. Conservation of all species of the organisms on the earth.

U.N. Conference on Human Environment for Stockholm declaration in year 1972 from 5th June to 16th June held in Stockholm. The following recommendations were made by the conference.

(i) Every human being has the right to live in healthy environment and his fundamental duty is to protect the environment and its surroundings.

(ii) Ecological balance should be maintained.

(iii) For the protection of the society the wild life should be properly managed.

(iv) Natural renewable resources should be maintained on the earth.

(v) No interference in geological balance, or over exploitation of natural resource as they lead to pollution.

(vi) Waste and pollutants should not be discharged directly into oceans.

(vii) Science and technology should be used for the identification of various disasters of environmental pollutants.

(viii) Every country must take into account that any work done in their country should not effect the other country.

(ix) Environment effects the economic development and quality of living of the human beings.

(x) Every one should be given education about the ill effect of deterioration of environment.

The late Prime Minister, Smt. Indira Gandhi played a vital role in the above conference. She had taken initiative and come forward with a definite programme on environmental conservation. She made 42nd amendment in the constitution. According to this amendment, the problems regarding forests, wild life and environment were considered. Laws have been made and fundamental duties imposed on every citizen of this country.

Under the auspices of the International Union of Conservation of Nature and Natural Resources (IUCN), World Wildlife Fund (WWF) and UNEP, a World Conservation strategy was prepared and released for adoption and implementation in India on March 5,1980. The WCS is to ensure the management of human use of biosphere in a manner which may yield greater sustainable benefit to the present generation while maintaining its potential to meet the needs and aspirations of future generation.

Keeping in view the need for environmental protection, the Indian Parliament passed the water (Prevention and Control of Pollution) Act, 1974 which became effective from 23 March 1974. The Central Board for Prevention and Control of Water Pollution (CBPCW) was constituted in Sept. 1974. The Air (Prevention and Control of Pollution) Act 1981 was promulgated on 16 May, 1981. The Environment (Protection) Act 1986, was promulgated on 23 May, 1986.

The main instruments presently for control of air, water and noise pollution in country are:

(1) The water (Prevention and Control of Pollution) Act 1974.

(2) The Air (Prevention and Control of Pollution) Act 1981.

(3) The Environment (Protection) Act 1986.

(4) The Motor Vehicles Act, 1988.

Besides there are other acts to conserve our environment.

Factories Act (1948) — Amended in (1987): In Section 12 of our constitution it is stated that it is the duty of every industrialist to properly treat and manage the waste according to the plans laid down by the state government.

(i) In Section 11 of the act, it is given that it is the duty of the factory owner to check the factory is free from any effluent.

(ii) Any effluent producing factory should be away from human habitat.

(iii) Only limited amount of harmful waste should be emitted.

(iv) Machinery should be properly maintained and should be periodically checked.

Environment (Conservation) Act 1980 — (Amended 1988): According to this law:

1. Natural forests on the earth cannot be changed into other kind of plantation without prior permission of the Government.

2. If any one converted forest area for development project, the same area is to be planted by him.

3. Forest planning and management should be stressed.
4. No area should be deforested with the purpose of reforesting it.
5. There should be controlled grazing.
6. A forestation near the hills and slopes of the hills.
7. Encouragement to the tribal community.
8. Penalties for disobeying laws.

Forest (Conservation) Act, 1980—(Amended 1988): National Forest Policy 1952 enunciated that one third of the geographic area of the country should be under forests. However there had been continuous deforestation in the country for various reasons, and it is estimated that 4.23 8 Mha of forest land was officially diverted to non-forest purposes between 1951-52 and 1979-80. With a view to conserve forests, Govt. of India could enact the Forest (Conservation) Act, 1980.

1. Act was enacted with a view to check indiscriminate dereservation and diversion of forest land to non-forest purposes.
2. Under this act prior approval of Central Government is required before any reserved forest is declared as dereserved, or forest land is diverted to non-forest purposes.
3. If diversion is permitted, compensatory afforestation is insisted upon and other suitable conditions imposed.
4. Where non-forest lands are available compensatory afforestation be raised over equivalent area of non-forest land.
5. Where non-forest lands are not available, compensatory plantation be raised over degraded forests twice in extent to the area being diverted.
6. A control should be exerted over shifting cultivation and encroachments.

7. Grazing problems of the area should be studied and appropriate measure be adopted.
8. All forest working plans should stress conservation and have multi-disciplinary approach.
9. All critical areas in the hills, catchment areas, steep slopes and other parts under erosion must be protected and quickly afforested.

Six regional offices have been set up to monitor the conditions of forests and to find out the steps for their conservation. There are located at Bangalore, Shillong, Bhopal, Bhubaneshwar, Lucknow and Chandigarh.

However there are certain limitation to this Act.

(i) Since the prices of timber at rise therefore stealing and smuggling of finest timbers have increased.

(ii) The people have started encroachment in the forest area.

(iii) Since the number of grazing animals is on increase they usually roam in restricted areas.

(iv) The forests have been destructed for construction of roads, water supply schemes etc.

(v) It is very difficult in forests to apprehend the offenders.

Wildlife (Protection) Act, 1972—(Amended 1991): Wild life includes both plants and animals, is a living component of nature. It is must to maintain ecological balance, to prevent soil erosion and to obtain products which are economically important such as wood, medicines and drugs, resins, gums, fodder etc.

A number of wild life acts have been made from time to time, by state as well as Union Government for wildlife conservation such as: (a) Madras Wild Elephant Preservation Act, 1873 (b) All-India Elephant Preservation Act, 1879 (c) The

Wild Birds and Animals Protection Act, 1912 (d) Bengal Rhinoceros Preservation Act, 1932. (e) Assam Rhinoceros Preservation Act, 1954 (f) Wild Life (Protection) Act, 1972.

Indian Board of Wildlife (IBWL) is the main advisory body of the Govt. of India. It was first constituted in 1952 as an advisory body under the name Central Board of Wildlife. Later it was renamed as IBWL. At its first meeting, the board made a recommendation for unified legislation for wildlife conservation in India.

The Wildlife (Protection) Act was enacted in 1972 which was amended in 1991, which has been adopted by all states. The Indian Board of Wildlife which act under this act is chaired by Prime Minister. IBWL has asked both B.S.I. (Botanical Survey of India and Z.S.I. (Zoological Survey of India) to prepare a list of threatened species (which is likely to become extinct) of both plants and animals. These species are recorded in Red Data Book of IUCN (International Union for Conservation of Nature and Natural Resources).

At the 15th meeting of the IBWL held on 1st Oct. 1982, the then Prime Minister, Late Smt. Indira Gandhi gave a 12 Point strategy for an Action Plan for the conservation of wildlife in India. This included the establishment of a net work of scientifically managed protected areas including National Parks, Sanctuaries, Biosphere reserves.

National Parks are the areas maintained by Govt. where cultivation, grazing, forestry and habitat manipulations are not allowed.

Sanctuaries are tracts of land with or without cakes where hunting of animals is not allowed but other activities such as tilling of land, collection of forest products etc. are allowed.

Biosphere reserves are multipurpose protected areas where wildlife, tribals, domesticated plants and animals are allowed to live in harmony.

The main recommendations of the act are:

(i) Trapping and hunting of wild life should be prohibited.

(ii) Conservation and effective control of poaching wildlife.

(iii) Setting of national parks, sanctuaries and zoological gardens.

(iv) Ban on the export of living animals, skins, fur, leather and other wild life products.

(v) Human activity is not allowed in the core zone of biosphere reserve and national parks.

(vi) Uprooting, picking and sale of endangered plants not allowed.

(vii) No arm licence is to be issued in an area within 10 km of a Sanctuary.

Limitations

(a) Less attention is given to endangered plant species

(b) Exploitation of plant species continue under the garb of traditional medicine as the plants of medicinal importance are not cultivated.

(c) Wildlife authorities take a lot of time to decide whether an animal is dangerous and by the time it is decided it would have killed a number of domestic animals.

(d) Stress is given only on conservation of one or two species not all.

Environment (Protection) Act, 1986: The Act was promulgated to provide for the protection and improvement of environment and matters connected therewith. As per Environment (Protection) Act 'environment' includes water air and land and the inter relationships which exists among and between water, air and land and human beings, other living creatures, plants, micro-organisms and property;

"environment pollutant" means any solid liquid or gaseous substance present in such concentration as may be injurious to environment; "environment pollution" means the presence in the environment of any environmental pollutant, and "hazardous substance" means any substance which is liable to cause harm to human beings and other living creatures. The Act provides general power to the central government to take all necessary measures for the purpose of:

(a) Protecting and improving the quality of the environment.

(b) Preventing, controlling and abating environmental pollution.

The functions of Central Govt., are:

1. Planning and execution of nation wide programme for the prevention, control and abatement of environmental pollution.
2. Laying down standards for quality of environment in its various aspects.
3. Laying down standard for emission or discharge of environmental pollutants from various sources.
4. Carry out and sponsor investigations and research.
5. Collection and dissemination of information on environmental pollution.
6. Examination of such manufacturing processes materials and substances which are likely to cause environmental pollution.
7. Demarcation of areas in which certain industries/ processes/operations shall be not or shall be carried out subject to certain safeguards.
8. Laying down procedures and safeguards for tackling of restricted prohibited and hazardous substances.

9. Preparation of manuals, codes and guides regarding prevention, control and abatement of environmental pollution.

10. Establishment or recognition of environmental laboratories.

Air (Prevention and Control of Pollution) Act, 1981 (Amended 1987): Air (Prevention and Control of Pollution) Act was promulgated under article 253 of constitution for prevention control and abatement of air pollution by creating central and state boards.

Central Board exercises powers and performs the functions of State Board for Union Territories either directly or by delegating powers.

The various functions of Central Board are:

(i) To provide guidance and technical assistance to state boards and the industries.

(ii) To advise both Central Government and State Government regarding improvement in methods related to check air pollution.

(iii) To provide training to the persons involved in the field of air pollution.

(iv) To set up laboratories to check all kinds of samples.

(v) To educate people with the assistance of mass media.

The various functions of State Board are:

(i) To advise State Government to combat with the problem of air pollution.

(ii) To collect information regarding causes, prevention and control of air pollution.

(iii) To lay down standards for air quality.

(iv) To inspect air quality periodically to check air pollution.

As per Air (Prevention and Control of Pollution) Act both Central and State Boards have been given certain powers to meet the consequences due to air pollution.

(a) To declare any area within the state as air polluting area.

(b) Before setting up any unit every industrial establishment has to take clearance from the board. The board has got power to refuse or grant consent depending upon whether the industry fulfils the specification laid down by the board.

(c) The board has got power to stop industrial operations in air pollution control areas.

(d) The board collects the samples of air or emission for analysis.

(e) The board has got power to cancel the consent given to industry at any time.

(f) The board officials have got power to go to any industry at any time to check compliance with requirements of act.

(g) The board has got power to prosecute any defaulter.

The Air (Prevention and Control of Pollution) Act, 1981 was amended in 1987 to remove the difficulties encountered during implementation, to confer more powers on the implementing agencies and to impose more stringent penalties for violation of the provisions of the Act. The main point was also to amend the definition of air pollutants to include noise also.

The Water (Prevention and Control of Pollution) Act, 1974 (Amended 1988): The Water (Prevention and Control of Pollution) Act was enacted under article 252 (1) of Constitution as measure to prevent and control water pollution. The pollution under the act includes (a) any contamination in water (b) any physical, chemical or biological alteration in water (c) direct or indirect discharge of sewage, trade effluent or any other substance in water, (d) the discharge causing nuisance or harmful to public health. Central and State boards were created to meet the problem regarding water pollution.

The water (Prevention and Control of Pollution) Act, 1974 was amended in 1988. An important amendment was to rename the Central State Boards for Prevention and Control of Water Pollution as Central/ State Pollution Control Boards as Boards also deal with air pollution. More powers were given to CPCB.

Functions of Central Board

(i) To advise Central Government and State Governments about issues related to water pollution.

(ii) To provide guidance and training to persons in the field of water pollution.

(iii) To educate people through mass media.

(iv) To provide technical assistance and guidance to state boards and industries.

(v) To set up laboratories to analyse samples.

Functions of State Board

(i) To advise state Government regarding issues related to water pollution.

(ii) To seek guidance and training of persons connected with prevention and control of water pollution.

(iii) To collect information regarding causes, prevention and control of water pollution.

(iv) To organise programmes to control water pollution.

(v) To find out recent methods for disposal of treated sewage and effluents.

Water (Prevention and Control of Pollution) Act has provided different powers to both Central/State Pollution Control Boards:

1. It is the duty of each industrial establishment to take consent from state board on a specific application about the methods of treatment of disposal of sewage and other industrial effluent be discharged in the water body.
2. Board has got power to take sample of any sewage, industrial effluent from any water body.
3. Board has power to enter any industrial premises for inspection.
4. Board has got power to move any application to court after an offence is detected.
5. Board can ask closure of any unit any time It can stop water supply, electricity to the units of defaulters.

The Motor Vehicle Act, 1938 (Amended 1988): This Act has come into force from 1st July 1989. The salient features of the provisions of the Act, besides other things also include the requirements to be observed in matter of vehicle fitness. Some of the features are as follows:

1. For registration of a new vehicle, the same is to be produced before the registering authority for inspection to satisfy that vehicle is fit for registration. The certificate of fitness is to be issued for a new vehicle for a period of two years. Thereafter, the renewal shall be for period of one year till the vehicle attains the age of 10 years.
2. Every vehicle is required to meet the safety stands of components.
3. Each vehicle must have all the components of the standard laid down by Bureau of Indian Standards.

4. The vehicle must have tested by VRDE, Ahmadnagar, ARAI, Pune or CMITI, Budni or any other agency specified by the Central Government.
5. Manufacturer is to ensure that the vehicle does not cause pollution *i.e.* It is pollution free.
6. The horn to be used is to be in accordance with the approved specifications of Bureau of Indian Standards.
7. The vehicle should be fitted with tune up and catalytic converter.
8. Each vehicle is required to obtain "Pollution under control" certificate every 6 month.
9. All future vehicles are to have engines based on unleaded petrol.
10. Transportation of Goods of dangerous or hazardous nature to human life has been regulated under the Act. The vehicle carrying such goods should have prescribed labels indicating the nature of hazardous goods. The package containing such goods should also bear specific labels.

Warming Globally

Global warming or green house effect is the warming of the earth due to emission of the harmful gases from the earth, which forms the blanket over the earth and do not allow the gases to move upward to the atmosphere. Since CO_2 is confined exclusively to the troposphere, its higher concentration may act a serious pollutant. Under normal conditions (with normal CO_2 concentration).

The temperature at the surface of the earth is maintained by energy balance of the sun rays that strike the planet and heat that is radiated back into the space. However, when there is an increase in CO_2 concentration, the thick layer of this gas prevents the heat from being re-radiated out. This thick carbon dioxide layer thus functions like the glass panels of a green house (or the glass windows of a motor car), allowing the sunlight to filter through but preventing the heat from being re-radiated out in outer space.

This is the so called green house effect. Thus most heat is absorbed by CO_2 layer and water vapours in the atmosphere, which adds to the heat that is already present. The net result is the heating up of earth's atmosphere. Thus increasing CO_2 levels tend to warm the air in lower layers of atmosphere on a global scale. Other gases of the green house effect on earth are methane, chlorofluorocarbons and nitrogen oxides. These are called green house gases. These gases are emitted to the atmosphere through burning of fossil fuels like coal, wood, oils etc.

It is estimated that more than 18 x 1012 tonnes of CO_2 is being produced annually from the fossil fuels only CO_2 concentration has increased from 280 ppm in 1800 to 359 ppm in 1994 due to increasing use of fossil fuels and decreasing forest cover. More methane is being released from paddy fields due to intensive cultivation, cattle due to large population and escape from gas plants. Combustion of fossil fuels (along with excessive use of nitrogen fertilizers) is also the causative agent for higher concentration of nitrogen oxides.

They react with hydrocarbons to form ozone. As there is an increase in the release of CO_2 it is estimated that temperature increases after every 100 years from 0.3° to 0.7°C. If these gases are not minimised in the year 2030 the concentration of CO_2 will increase twice the present range and the temperature of the earth surface will increase by 1.5°C to 4.5°C.

Effects

1. Global warming shall cause partial melting of polar and alpine ice caps that would result in raising sea level. Already the sea level has risen 15 cm in the past

century. If the trend continues, there is a danger of submersion of large area *e.g.* Maldives, and six other Coral atoll countries, several thousand islands.

2. A continuous increase of green house gases, leads to the rise in mean global temperature which affects the climate leading to change in rainfall pattern, conversion of fertile lands into deserts, shortage of water due to evaporation.
3. Due to change in weather conditions, the production of crops is also affected and reduces the yield of grains like rice, maize, wheat etc.
4. Chances of hurricanes, cyclones and floods will increase.
5. Green house effects disturb the food chain and destroy many cold habitat species.
6. The forests present in the middle latitudes will be wiped out.
7. Several lakes would dry up.

So, UNEP has appropriately chosen the slogan "Global Warming: Global Warming" and since 1989 5th June is celebrated as World Environment Day.

These problems can be overcome by awareness among people and plantation of trees and use of non-conventional sources of energy.

10

Analytical Variants

Once the experimental results are obtained they have to be analysed and interpreted. In experimental situations we may have large number of treatments. Our interest will be to test whether all the treatments have the same effect. In other words, our intention will be to test the null hypothesis, $H_0 : \mu_1 = \mu_2 = \mu_3 = ... = \mu_t$, against the alternative hypothesis, $H_1 : \mu_1 \neq \mu_2 \neq \mu_3 \neq ...... \neq \mu_t$. For this purpose we may use Student's *t*-test by taking all possible combinations. But it is a tedious procedure and theoretically not valid. The appropriate method for such tests is the *analysis of variance.*

The analysis of variance can be regarded as a special development of regression analysis. The fundamental principles of analysis of variance and its most important techniques were developed by R.A. Fisher. In the early stages, the analysis of variance technique was primarily applied to agricultural and biological experiments. Now it is used in every field of science.

Analysis of Variance

The analysis of variance is the systematic algebraic procedure of decomposing the overall variation in the responses observed in an experiment into different components. Each component is attributed to an identifiable cause or source of variation. The structure of these component parts is determined by the design of experiments.

Suppose that we have a factor *T* with *t* levels (that is, *t* treatments) with equal replication, *r* and their responses, *Y*. The results may be tabulated as shown in the following table.

The Arrangement of Data

	Treatments				
	1	*2*	*i*	*t*	
	Y_{11}	Y_{21}	Y_{i1}	Y_{t1}	
	Y_{12}	Y_{22}	Y_{i2}	Y_{t2}	
	..	..	..	..	
	Y_{1j}	Y_{2j}	Y_{ij}	Y_{tj}	
	..	..	..	..	
	Y_{1r}	Y_{2R}	Y_{ir}	Y_{tr}	
Total:	$Y_{1\cdot}$	$Y_{2.}$	$Y_{i\cdot}$	$Y_{t.}$	$Y..$ = Grand Total
Mean:	Y_1	Y_2	Y_i	Y_t	

The response of the jth individual unit in the ith treatment may be represented by the equation.

$$Y_{ij} = Y_i + e_{ij} = \text{i}^{\text{th}} \text{ treatment mean + random deviation}$$

$$= \mu + (Y_i - \mu) + (Y_{ij} - Y_i)$$

or, $$(Y_{ij} - \mu) = (Y_i - \mu) + (Y_{ij} - Y_i)$$

That is, the variation of an individual response from the overall mean is the sum of two components. One is the variation in the means of the treatments, Y_i from the overall mean, μ.

The other is the variation within each of the treatments. The variation within each treatment is not accounted for by the treatment, and hence it is the error term.

The significance of the difference $(Y_i - \mu)$ is based on the magnitude of the components $(Y_i - \mu)$ compared with $(Y_{ij} - Y_i)$. The statistical test of the significance involves the use of the following identity.

$$\Sigma\Sigma(Y_{ij}-\mu)^2 = \Sigma\Sigma\left[(Y_i-\mu)+(Y_{ij}-Y_i)\right]^2$$

$$= \Sigma\Sigma(Y_i-\mu)^2 + \Sigma\Sigma(Y_{ij}-Y_i)^2 + 2\Sigma\Sigma(Y_i-\mu)(Y_{ij}-Y_i)$$

$$= \Sigma\Sigma(Y_i-\mu)^2 + \Sigma\Sigma(Y_{ij}-Y_i)^2$$

The last term will vanish because $\Sigma(Y_{ij}-Y_i)=0$

Thus we have,

Total sum of squares = treatment sum of squares + error sum of square,

This can be abbreviated as,

SS(Y) = SS(T) + SS(E)

The total variation can thus be partitioned into two components, *i.e.*, variation between treatments and variation within treatments.

Before the magnitude of the two components can be compared, they must be divided by the number of degrees of freedom.

For the component "between treatments", the degrees of freedom is t - 1, and that for the "within treatments" is $(tr - t) = (n-t)$ or $t(r - 1)$.

The quantity *SS(T)/t* - 1 is known as the treatment mean square and written as *MS(T)*. The quantity *SS(E)/n -t* is known as the error mean square and written as *MS(E)*. In general, treatment mean square is denoted by s_t^2 and error mean square by s_e^2. If samples were drawn randomly, s_t^2 and s_e^2 are the independent estimates of the same quantity, σ^2, population variance of Y. Therefore, the ratio between two estimates,

$$F = \frac{MS(T)}{MS(E)} = \frac{s_t^2}{s_e^2}$$

follows the F-distribution with $(t - 1)$ and $(n - t)$ degrees of freedom. The significance of F can be determined in the usual way by using the Tables of F. Thus, the F-statistic may be used to test hypotheses about the equality of the population means of the treatments, that is, to test $H_0 : \mu_1 = \mu_2 = ... = \mu_t$. If F is significant the treatment means will be significantly different. The analysis of variance concept can be understood by the following examples.

Example

An experiment was conducted to study the effect of four levels of nitrogen on yield of paddy. The four levels of nitrogen were,

No application of nitrogen (t_1),

50 Kg nitrogen per hectare (t_2),

100 Kg nitrogen per hectare (t_3), and

200 Kg nitrogen per hectare (t_4).

The yields of grain in kg per plot are shown in the following table.

Grain Yield of Paddy in Kg/plot

t_1	t_2	t_3	t_4
52	58	62	66
50	52	58	69

Contd..

	t_1	t_2	t_3	t_4	
	56	60	65	70	
	54	57	58	67	Grand
	53	56	61	66	Total
Total	265	283	304	338	1190

Computational steps:

Step 1: Compute the correction factor (CF)

$$CF = \frac{(\text{Grand total})^2}{n} = \frac{Y_{..}^2}{n}$$

$$= \frac{1190^2}{20} = \frac{1416100}{20}$$

$$= 70805$$

Step 2: Compute the total sum of squares

$$\text{Total} \quad SS = \Sigma\left(Y_{ij}^2 - \mu\right)^2 = \Sigma Y_{ij}^2 - \frac{Y_{..}^2}{n}$$

$$= \Sigma Y_{ij}^2 - CF$$

$$= (52)^2 + (50)^2 + ... + (66)^2 - CF$$

$$= (2704 + 2500 + ... + 4356) - CF$$

$$= 71498 - 70805 = 693$$

Step 3: Compute the treatment (or group) sum of squares.

$$\text{Treatment SS} = \Sigma\left(\text{Y}_{\text{i}} - \mu\right)^2$$

$$= \Sigma \frac{Y_{i.}^2}{r_i} - \frac{Y_{..}^2}{n}$$

$$= \frac{(265)^2}{5} + \frac{(283)^2}{5} + \frac{(304)^2}{5} + \frac{(338)^2}{5} - CF$$

$$= \frac{1}{5}\left(265^2 + 283^2 + 304^2 + 338^2\right) - CF$$
$$= \frac{1}{5}(356974) - 70805$$
$$= 71394.8 - 70805$$
$$= 589.8$$

Step 4: Compute the error sum of squares (within treatment *SS*).

Error $SS = \Sigma\left(Y_{ij} - Y_i\right)^2$

$$= \Sigma Y_{ij}^2 - \frac{(\Sigma Y_i)^2}{r_i}$$
$$= \left(52^2 + 50^2 + 56^2 + 54^2 + 53^2\right) - \frac{265^2}{5}$$
$$+ \left(58^2 + 52^2 + 60^2 + 57^2 + 56^2\right) - \frac{283^2}{5}$$
$$+ \left(62^2 + 58^2 + 65^2 + 58^2 + 61^2\right) - \frac{304^2}{5}$$
$$+ \left(66^2 + 69^2 + 70^2 + 67^2 + 66^2\right) - \frac{338^2}{5}$$
$$= \left(14065 - \frac{70225}{5}\right) + \left(16053 - \frac{80089}{5}\right)$$
$$+ \left(18518 - \frac{92416}{5}\right) + \left(22862 - \frac{114244}{5}\right)$$
$$= (14065 - 14045) + (16053 - 16017.8)$$
$$+ (18518 - 18483.2) + (22862 - 22848.8)$$

$$= 20.0 + 35.2 + 34.8 + 13.2$$
$$= 103.2$$

Alternatively,

Error *SS* = total *SS* - treatment *SS*

= 693.0 - 589.8 = 103.2

These results can be put in the form of a table. This table is known as analysis of variance (ANOVA) table. For our example, the ANOVA table is as follows:

Analysis of Variance for the Data in Table

Sources of variation	*degrees of freedom (df)*	*sum of squares (SS)*	*Mean squares (MS)*	*F*
Between treatments	(4 - 1 =) 3	589.8	196.60	30.481**
Error	(19-3=) 16	103.2	6.45	-
Total	(20-1=) 19	693.0	-	-

In general, the ANOVA table for one-way classification will be of the form as given in Table.

General form of the analysis of variance table for one-way classification:

Sources	*df*	*SS*	*MS*	*F*
Between treatments	t -1	*SS(T)*	$\frac{SS(T)}{t-1} = s_i^2$	s_t^2 / s_e^2
Error	n - t	*SS(E)*	$\frac{SS(E)}{n-t} = s_e^2$	
Total	n - 1	*SS(Y)*	-	-

For our example, table value of F for (3, 16) *d.f.* and at 1 percent level of significance is 5.29. Since the observed F-value is larger than the table value, it is significant at 1 percent level. This is indicated by two asterisks. If it is significant at 5 percent level it will be indicated by one asterisk.

The columns in the ANOVA table will be same for all types of designs. However, the rows representing the sources of variation will differ according to the type of experimental design. The form of the analysis of variance table and the computational formulae for the sums of squares used in the analysis of variance are given under each design of experiment in the subsequent chapters.

Assessment of Variance Model: We have seen that the jth individual unit in the ith treatment may be represented by the equation:

$$Y_{ij} = \mu + (Y_i - \mu) + (Y_{ij} - Y_i)$$

This equation can be written as

$$Y_{ij} = \mu + t_i + e_{ij}$$

where, $t_i = i^{th}$ treatment effect, and

e_{ij} = error term.

This means that any Y_{ij} is made up of overall mean + treatment effect + an error term. Since the effects are added in this model, it is known as a linear additive model. It is commonly known as *analysis of variance model.* Depending on the design, the analysis of variance model will take different forms. They are given under each design in later chapters.

Analysis of Variance: Its Two Types of Models: Usually the treatments in an experiment are fixed by the experimenter. They are not selected randomly from all possible treatments. For example, in a fertilizer experiment we may select three levels of nitrogen as 60, 80 and 100 Kg per hectare and test whether all these three levels of nitrogen have the same effect on yield of a crop. These three nitrogen levels were not selected randomly from all possible levels of nitrogen. They were determined by us as per our convenience. When the treatments are fixed like this, the analysis of variance model is known as *fixed effects model* or *Model I.*

There are instances in agricultural investigations where the treatments are not fixed like this. For example, we might be interested in the effect of geographic locations on the yield of a crop. It is possible that our concern might be with certain specific locations. In that case we would employ Model I. Instead, if our interest is to generalize to all possible locations from which they have been selected we would employ model II. As another example, suppose that we wish to determine the nitrogen content in paddy plants. We may select 5 hills at random from large number of hills.

From each selected hill three independent determination of the nitrogen content may be made. The results may then be generalised to all hills. Thus, when t treatments are selected at random from a population of T treatments, we have *random effects model* or *Model II*. It is also known as *variance-components model*.

Under model I, it is assumed that all treatments about which inferences are to be made are included in the experiment. If the experiments were to be replicated, the same set of treatments would be included in each of the replication. On the other hand, under Model II, a different random sample of t treatments would be included in each of the replication.

For both model I and model II, the equation is same. For example, the equation:

$$Y_{ij} = \mu + t_i + e_{ij}$$

is appropriate for both the models. However, the assumptions underlying these models are not same. In model II, the term t_i is a random variable. The distribution of t_i is assumed to be normal with mean 0 and variance σ_t^2. The variance 'between treatments' will be $\sigma^2 + \sigma_t^2$, where σ^2 is the error variance. The assumptions about the error term are, however, same in both the models.

The procedures will be same for both the models in initial significance tests. After the initial significance test, a Model I analysis of variance is completed by examining the data in greater details. In other words, the null hypothesis under model I is $H_0 : \mu_1 = \mu_2 = ... = \mu_t$. In model II, we will be interested in estimating the variance σ_t^2 rather than the individual t_i. The null hypothesis under model II is, therefore, $H_0 : \sigma_t^2 = 0$. The relative amounts of variation 'between treatments' $(\sigma^2 + \sigma_t^2)$ and 'within treatments' (σ^2) would guide us in designing further studies of this sort. If 'between treatments' variation is significantly more than 'within treatments' (error) variation, we would need more treatments and fewer replications per treatment. On the other hand, if 'between treatments' variation is not significantly more than 'within treatments' variation, we would need fewer treatments and more replication per treatment.

When more number of factors are used, we may have a mixed model. If all the levels of one factor are used while a random sample of the levels of another factor is used in an experiment, a mixed model results.

Model I is better understood than the other models discussed above. Hence, it is used very commonly. It is the most appropriate model for single factor experiments. However, in multi-factor experiments choice should be made between model I and model II.

The Theories

There are certain basic assumptions underlying the analysis of variance model. They are:

1. The effects of the different factors (treatments and environment) are additive.
2. The experimental errors are independent.
3. The experimental errors are distributed normally with mean 0 and common variance σ^2.

The usual interpretation of the analysis of variance is valid only when such assumptions are met. When the data deviate from these assumptions to lesser extent the analysis of variance technique will not be affected seriously.

However, larger deviations from these assumptions will affect both the level of significance and the sensitivity of *F* and *t* tests. When the level of significance is affected, the results may look more significant than they actually are. For example, the difference between the effects of two treatments may be significant only at 6 percent level. But we may declare that it is significant at 5 percent level. This will lead to invalid conclusions.

The meaning of the above assumptions, identification of failure of these assumptions and the remedial measures are described in the following sections.

The Process of Adding: When the effects of two factors are independent, the effects of one factor remain constant over all levels of the other factor. In that case, the effects of two factors are said to be additive. For example, consider the hypothetical data in the following table.

Numerical Example for Additivity

Treatment		*Replication*			*Replication effect*		
		I	II	III	II-I	III-I	III-II
1		20	25	35	5	15	10
2		30	35	45	5	15	10
3		40	45	55	5	15	10
Treatment	2-1	10	10	10			
effect	3-1	20	20	20			
	3-2	10	10	10			

From Table, it can be seen that the effect of treatment 2 (treatment 2 -treatment 1) is 10 for ill replications. Similarly, the effect of treatment 3 is 20 (treatment 3 - treatment 1) or 10 (treatment 3 - treatment 2) for all replications. In the same way, it can be found that the effect of replication II is 5 for all the treatments and the effect of replication III is 15 (replication III - replication I) or 10 (replication III - replication II) for all the treatments. Here the effects of treatments and replications are additive.

Let us now consider the hypothetical data in the table below.

Numerical Example for Non-additivity

Treatment		*Replication*			*Replication effect*(%)		
		I	II	III	II-I	III-I	III-II
1		20	30	40	50	100	33
2		30	45	60	50	100	33
3		40	60	80	50	100	33
Treatment	2-1	50	50	50			
effect (%)	3-1	100	100	100			
	3-2	33	33	33			

The percentage increases (effects) are computed as:

$$\frac{30-20}{20} \times 100, \frac{45-30}{30} \times 100, \frac{60-40}{40} \times 100. \text{ and so on.}$$

From Table, it can be observed that the effect of replication II is 50 percent for all treatments. The replication effect is, therefore, not constant over all treatments. The treatment effects are also not constant for all replications. The increase is by a percentage in this case.

When the effects are by certain percentages as in the example, they are said to be *multiplicative* or *non-additive*. If the analysis of variance model for additive effects is given by

$$Y_{ij} = \mu + t_i + r_j + e_{ij}$$

the model for multiplicative effects may be given by

$$Y_{ij} = \mu + t_i + r_j + (tr)_{ij} + e_{ij}$$

For the purpose of examining additivity we may use Tukey's test for non-additivity. For details about this test one can refer Snedecor and Cochran (1967).

In agricultural experiments, the presence of non-additive effects is common in fertilizer trials and insect and disease studies. In fertilizer trials, the plots with low level of available nitrogen in soil, for example, will benefit more from addition of nitrogen than plots with adequate level of available nitrogen. In studies on the incidence of insects or diseases, the incidence pattern will be in multiples of the initial incidence. This happens because of the nature of the growth of insect or disease organisms. In these situations, the assumption of additivity would not be correct.

When non-additive effects are present, the experimental error will be inflated. Consequently, the precision will be lost. In order to overcome this problem the multiplicative effect may be corrected by applying logarithmic transformation.

Factors of Normality

We know that many statistical tests are valid only when the data follow normal distribution. In some experimental situations the distribution of errors may not be exactly normal as assumed under the analysis of variance model. It may be skewed or it may follow a Poisson distribution or a binomial distribution. There are many methods to test the normality.

For testing the normality we need fairly large samples. With small samples it will be difficult to prove normality.

When extreme values are present in a treatment it is an indication of skewed distribution. If the ratio of largest value to smallest value is 2 or more, we may say that there are extreme values.

The nature of the functional relationship between the mean and variance may help in identifying the non-normal distribution. The functional relationship may be $\sigma^2 = \mu$. In such a case, the variance is proportional to the mean. This kind of relationship exists when the distribution is Poisson in form. We can also identify Poisson distribution by the nature of data. Often, counts of events having a small probability of occurrence follow Poisson distribution. For example, count data, such as the number of insects per plant, the number of insects caught in a trap, the number of infested plants per plot, the number of weeds per plot, the number of lesions per leaf, etc., usually will approximate Poisson distribution.

The functional relationship, $\sigma^2 = \mu(1-\mu)/n$ occurs when the distribution is binomial. When the observations are proportions or percentages like germination percentage of seed materials, percentage mortality of insects, percentage of infestation and percentage of barren tillers will approximate binomial distribution. In the usual notation of binomial distribution, $\mu = p$ and $1-\mu = q$.

The non-normal distribution of errors produce too many significant results. Also, there will be loss of efficiency in the estimate of the treatment means.

In order to normalise a non-normal distribution, we can use data transformation. Logarithmic transformation is used to normalise a skewed distribution; the square root transformation is used to normalise a Poisson distribution; and the angular transformation is used to normalise a binomial distribution.

Homogeneity of Variance

In the analysis of variance the usual estimate of error variance is the pooled variance. This pooling is valid only when the variances are homogeneous. That is why the assumption of common error variance is made in the analysis of variance.

The meaning of common error variance can be explained with an example. The error term for individual responses of each treatment can be written as

$$e_{ij} = Y_{ij} - Y_i$$

The mean and variance for these e_{ij} values can be computed for each treatment as given in table given below.

Numerical Example for Common Variance of Error Term

	Treatment			*e_{ij} for treatment*			*e_{ij}^2 for treatment*	
1	2	3	1	2	3	1	2	3
2	7	10	-1	1	0	1	1	0
4	8	12	1	2	2	1	4	4
3	4	8	0	-2	-2	0	4	4
1	5	9	-2	-1	-1	4	1	1
5	6	11	2	0	1	4	0	1
Total 15	30	50	0	0	0	10	10	10
Mean 3	6	10	0	0	0	2	2	2

From Table, it can be seen that the mean of error term for each treatment is 0 and the variance is 2.

In practice we may not get the variances exactly equal. However, they should be close to each other. The equality of many variances may be tested by using Barlett's or the test proposed by Hartley.

Hartley's test is simpler as compared to Bartlett's test. When the number of replications is same for all the treatments, the test statistic proposed by Hartley is given by

$$F_{\max} = \frac{\text{Largest variance of the } t - \text{treatment variances}}{\text{Smallest variance of } t - \text{treatment variances}}$$

Hartley has tabulated the sampling distribution of F_{max} statistic. Against the number of treatments *(t)* and the replication degrees of freedom (r - 1), the table value can be read. If the observed F_{max} value is greater than the table value for a specified level of significance, then the variances are declared as heterogeneous.

Yet another simple method for detecting the variance heterogeneity is to use the relationship between the range and the mean. The range tends to be proportional to the variance. Hence the range is used in place of variance. When the ranges and the means of treatments are plotted on a graph sheet, it will result in a scatter diagram.

Usually, heterogeneity of variance occurs when the variance is functionally related to the mean. We have seen some of the functional relationships between the variance and the mean in Section earlier. The functional relationship may also be of the forms, $\sigma^2 = C^2\mu$ and $\sigma^2 = C^2\mu^2$, where C is a constant. These types of variance heterogeneity are usually associated with non-normal distributions.

There is another type of variance heterogeneity caused by the nature of treatments. For example, plots treated with certain insecticides may have higher variation because the insect population is not uniform or due to non-uniform application of insecticides. The variances of varieties that are highly tolerant to moisture stress are expected to be small compared to that of highly susceptible varieties. The heterogeneity of variances may be there due to the genetic variability of the varieties tested. We know that the genetic variability in F_2 generation is much higher than that of the F_1 generation.

Once the variances are found to be heterogeneous, the simplest remedy is to omit the treatments with very large variance from the analysis. If there are too many treatments to be omitted, then this method is not considered satisfactory. Instead, we can use variance stabilising transformations depending on the functional relationship between the variance and the mean. Often the assumptions of homogeneous variance and normality are violated simultaneously. If the variance is not homogeneous then the distribution is not usually normal. Hence the transformations used to normalise the distribution, *viz.*, log, square root and angular transformations, will result in variance homogeneity.

When the variances are not functionally related to the means we may resort to *error partitioning* to handle data with heterogeneous variance. This method should be applied only after eliminating gross errors, if any.

For error partitioning we have to group the treatments with homogeneous variances. For example, suppose that the treatments are paddy varieties and that they are classified as long, medium and short duration varieties having equal variances within each group. Then the sums of squares for varieties and error will be partitioned as follows:

Varieties	Between groups
	Varieties within group 1
	Varieties within group 2
	Varieties within group 3

Error Replication x between groups

Replication x group 1

Replication x group 2

Replication x group 3

The *F*-values are then calculated as the ratio of the 'group mean square' and the corresponding 'error mean square'. For example,

$$F(\text{group 1}) = \frac{\text{Within group 1 MS}}{\text{Replication x Group 1 MS}}$$

Freedom of Error

If the error of an observation is not dependent upon that of another, it is said to be independent. Under certain conditions it is possible that the experimental errors are correlated. In field experiments, it is often found that adjacent plots give similar yields. If we allocate the same treatments to adjacent series of plots the errors tend to vary together. For example, consider the following layout:

Replication I	*A*	*B*	*C*	*D*	*E*
Replication II	*E*	*A*	*B*	*C*	*D*
Replication III	*D*	*E*	*A*	*B*	*C*

In this layout treatments *A* and *B* are adjacent to each other in all replications. The error for treatments *A* and *B* can be expected to be more related as compared to that between *A* and *D*.

Non-independence of errors tends to produce too many significant results in *F* and *t*-tests.

It is difficult to detect the presence of non-independence of errors by inspection of the data alone. However, by inspecting the layout the non-independence of errors may be detected. The remedy to overcome the lack of independence of errors is the proper randomization.

Data Transformation: We have seen that in actual experiments the ideal conditions for the analysis of variance may not be obtained. When one or more assumptions fail we are left with two alternatives. One is that a new model can be developed to which the data may conform.

The second is that certain corrections may be made in the data in such a manner that the corrected data meet the assumptions of the analysis of variance model. Development of a new model is not easy. But it is easy to make corrections in the data.

The corrections are done by means of converting the data from their original form to a new form. This conversion of data is known as *transformation* of data. Different types of transformations are available. The most commonly used transformations are log transformation, square root transformation and angular transformation.

Log Transformation: When the original observation Y is converted to log Y, the conversion is known as log transformation. Although log to any base can be used, log to base 10 is generally the easiest. If the observed value is 0, a constant value preferably 1 is added to avoid negative logarithms. When such constant is added, it is added to all the observations.

The log transformation is particularly effective in normalising positively skewed distributions. It is also used to achieve additivity.

An example for log transformation is given in the table below.

Observed values and their log transformed values.

Treatment	Original Values			Log Values		
	Replication			Replication		
	I	*II*	*III*	*I*	*II*	*III*
1	20	30	40	1.30	1.48	1.60
2	30	45	60	1.48	1.66	1.78
3	40	60	80	1.60	1.78	1.90

It can be verified that the treatment and replication effects are not additive in case of original observations. However, the treatment and replication effects are additive after log transformation.

Square Root Transformation: If the original observation Y is converted to a new value by taking its square root, it is known as square root transformation. It is used in case of count data. When some of the observed counts are numerically small, say less than 10, the more appropriate transformation is $\sqrt{Y+0.5}$. The transformation of the type $\sqrt{Y}+\sqrt{Y+1}$ is also used.

The square root transformation is used when the observations follow a Poisson distribution.

Angles Transformation: In case of proportions, derived from count data, the observed proportion p can be changed to a new form $sin^{-1}\sqrt{p}$. This type of transformation is known as angular transformation. It is also known as 'arcsine' transformation or 'inverse sine' transformation.

It may be noted that the angular transformation is not applicable to percentage data which are not derived from count data. For example, percentage of marks, percentage of profit, percentage of protein in rice, infection index, etc, can not be subjected to angular transformation.

The computational work can be reduced greatly by using ready made table of angular transformation. In the tables, the values of p are given in percentages.

The angular transformation is not good when $p = \frac{o}{n} = 0$ or $p = \frac{n}{n} = 1$. The transformation is improved by replacing $\frac{o}{n}$ with $\frac{1}{4n}$ and $\frac{n}{n}$ with $1 - \frac{1}{4n}$, where n is the total number of units under observation.

When all the observed proportions lie between 30% and 70%, the angular transformation need not be used. This is because of the reason that the transformation in that range of values will not alter the conclusions.

The angular transformation is used to normalise the binomial distribution, especially when the observed proportions are in the range of 0 to 30% or 70 to 100%.

All the above transformations are used mainly to stabilise the variances. For the sake of convenience the situations in which the various transformations are used are summarized below:

Observed distribution	*Functional relationship between variance and mean*	*Transformation*
Poisson	$\sigma^2 = \mu$	Square root: $\sqrt{Y+0.5}$
Empirical	$\sigma^2 = C^2\mu$	Square root
Binomial	$\sigma^2 = \mu(1-\mu)/n$	Angular: $sin^{-1}\sqrt{p}$
Empirical	$\sigma^2 = C^2\mu^2$	Logarithm: Log Y

In making the choice of a transformation, we have three alternative procedures. The nature of data itself may suggest the appropriate transformation. As a second method, the graph showing the relationship between the range and the mean may be used. The third alternative is to use Tukey's test of additivity.

Once the transformation has been made, the analysis is carried out with the transformed data. The conclusions are drawn from such analysis. However, while presenting the results, the means and their standard errors are transformed back into original units.

While transforming back into the original units, some corrections have to be made. In case of log transformed data, if the mean value is X, the mean value in the original units will be:

$$Y = \text{antilog}\left[(X + 1.15V(X))\right]$$

instead of $\quad Y = \text{antilog}\ X.$

If the square root transformation had been used, then

$$Y = (X + V(X))^2$$

instead of $\quad Y = \bar{X}^2$

In the above formulae $V(X)$ is the variance of the mean X.

Exhibition of Anova Result: We have seen that the null hypothesis of equal treatment means, that is, $\mu_1 = \mu_2 = ... = \mu_t$ can be tested using the analysis of variance technique. The test statistic in this case is F which is the ratio of the treatment mean square to the error mean square. The calculated F value is then compared with the table value of F. The table value is read against the specified level of significance and treatment and error degrees of freedom.

If the calculated value of F is greater than the table value, then F is significant; otherwise F is not significant. If F is significant at 5% level of significance, it is indicated by placing one asterisk(*) on the computed F-value. If it is significant at

1% level of significance, two asterisks (**) are placed on the computed *F*-value.

A non-significant *F* may result either due to small treatment difference or a very large experimental error or both. It does not mean always that all the treatments have the same effect. When the experimental error is large, it is an indication of the failure of the experiment to detect treatment differences. In order to find out the reliability of the experiment, the coefficient of variation *(CV)* is used. It is computed as

$$CV = \frac{\sqrt{\text{Error } MS}}{\text{Overall mean}} \times 100.$$

If the *CV* is 20% or less, it is an indication of better precision of the experiment. When the *CV* is more than 20%, the experiment may be repeated and efforts made to reduce the experimental error.

In case of significant *F*, the null hypothesis is rejected. Then the problem is to know which of the treatment means are significantly different. Many test procedures are available for this purpose. The most commonly used tests are the *least significant difference (LSD)* and the *Duncan's multiple range test (DMRT)*. The least significant difference is also known as *critical difference (CD)*.

A Great Difference

The CD is a form of *t* test. Its formula is given by

$CD = t \cdot SE(d)$

where,

$$SE(d) = \sqrt{EMS\left(\frac{1}{r_i} + \frac{1}{r_j}\right)}$$

In case of equal replications $SE(d) = \sqrt{\frac{2EMS}{r}}$

In the formula, t is the critical (Table) value of t for a specified level of significance and error degrees of freedom; and r_i and r_j are the number of

replications for r^{th} and j^{th} treatments, respectively. This formula is derived from the formula for t to test the significance of the difference between two means:

$$t = \frac{Y_i - Y_j}{\sqrt{s^2\left(\frac{1}{r_i} + \frac{1}{r_j}\right)}}$$

Two treatments are declared significantly different at a specified level of significance if their difference exceeds the calculated CD value; otherwise they are not significantly different.

The advantage of CD is that it is easy to calculate and it provides a single value for testing all differences. However, there are certain limitations in the use of CD. It should be used only if F is significant. It should be used only if the pair comparisons are planned before the data are examined. Comparison of the control with each of the other treatment is a common example for planned pair comparison. If the comparison is selected after the treatment means are observed, a certain number of differences in a set of t treatments will be large owing to sampling variation.

Multiple Range Test of Duncan: In a set of t treatments if comparison of all possible pairs of treatment means is required we can use Duncan's multiple range test. The DMRT can be used irrespective of whether F is significant or not. It involves number of steps which are as follows:

Step 1: Arrange the treatments in decreasing order, that is, according to their ranks.

Step 2: Calculate the standard error of mean as

$$SE(Y) = \sqrt{\frac{s_e^2}{r}} = \sqrt{\frac{EMS}{r}}$$

Step 3: From statistical Tables write the significant studentized ranges *(r_p)* for p = 2, 3,..... t treatments and error degrees of freedom.

Step 4: Calculate the shortest significant ranges *(R_p)*, where

$$R_p = r_p \,.\, SE\ (Y)$$

Step 5: From the largest mean subtract the R_p for largest p. Declare all the means less than this value as significantly different from the largest mean. For the remaining treatments whose values are larger than the difference (Largest mean-larges *R_p)*, compare the differences with appropriate R_p value. If two treatments are remaining compare with *R_2;* if three are remaining compare with *R_3;* and so on.

***Step 6*:** From the second largest mean subtract the second largest R_p and compare the treatments as in step 5.

Continue this process till all treatments are covered.

Step 7: Present the results by using either the line notation or the alphabet notation to indicate which treatments are on par and which are significantly different.

Example

From an experiment conducted in Randomized Block Design, the following results were obtained: Error mean square = 0.00193; error degrees of freedom = 12; number of replications = 4; mean of treatment 1, $(\bar{T}_1) = 0.36$; $(\bar{T}_2) = 0.16$; $(\bar{T}_3) = 0.21$; $(\bar{T}_4) = 0.16$ and $(\bar{T}_5) = 0.27$.

$$SE(d) = \sqrt{\frac{2EMS}{r}} = \sqrt{\frac{2(0.00193)}{4}} = 0.0311$$

For 12 degrees of freedom and 5 percent level of significance the table value of t is 2.18. Hence,

$CD = t \cdot SE(d)$

$= (2.18)\ (0.0311) = 0.0678$

The arrangement of treatments according to their ranks is

$T_1 \qquad T_5 \overline{\quad T_3 \quad T_2 \quad} T_4$

The difference between the means of T_1 and T_5 is 0.36 - 0.27 = 0.09. It is greater than the CD value. Hence, T_1 and T_5 are significantly different. Since other values are less than T_5 value they are also significantly different from T_1.

The difference between the means of T_5 and T_3 is 0.27 - 0.21 = 0.06. It is less than the CD value. Hence, they are not significantly different. To indicate this a connecting line (bar) is drawn on the top (or bottom) of the treatments. It can be verified that T_5 is significantly different from T_2 and T_4; and T_3, T_2 and T_4 are not significantly different.

The presentation of results as given above is known as *bar chart.* If the treatments are not significantly different they are said to be *on par.*

The DMRT can also be applied for the above data.

Treatment	Mean
T_1	0.36
T_5	0.27
T_3	0.21
T_2	0.16
T_4	0.16

$$SE(Y) = \sqrt{\frac{EMS}{r}}\sqrt{\frac{0.00193}{4}} = 0.0220$$

p	R_p	$R_p = r_p SE(Y)$
2	3.08	0.068
3	3.77	0.083
4	4.20	0.092
5	4.51	0.099

The largest mean - the largest R_p = 0.36 - 0.099 = 0.261. Except the mean of T_5 all others are less than 0.261.

Hence, T_1 is declared as significantly different from T_3, T_2 and T_4. Now, T_1 and T_5 are remaining and their difference is

$T_1 - T_5 = 0.36 - 0.27 = 0.09$

Since two treatments are remaining this value has to be compared with R_2. The difference is more than R_2 value (0.068). Hence, T_1 and T_5 are declared significantly different. The second largest mean - the second largest $R_p = T_5 - R_4$

$= 0.27 - 0.092 = 0.178$

The means of T_2 and T_4 are less than 0.178. Hence, T_5 is significantly different from T_2 and T_4.

The remaining two treatments are T_5 and T_3. Their difference is 0.27 - 0.21 = 0.06. This is less than R_2 value. Hence they are not significantly different. In other words they are on par. To indicate this a vertical line is drawn connecting them.

The third largest mean-the third largest R_p = 0.21 - 0.083 = 0.127. No treatment mean is less than this value. Three treatments are remaining now.

Hence for comparing the difference between their means, R_3 value (0.083) is taken. It can be seen that $T_3 - T_2$, $T_3 - T_4$ are less than 0.083. Hence all the three are on par.

Instead of using lines alphabets can be used as follows:

T_1	*a*
T_5	*b*
T_3	*bc*
T_2	*c*
T_4	*c*

When two treatments have a letter in common they are said to be on par. If the treatments have different letters they are declared as significantly different.

Class Comparison: In earlier Sub-section we have considered the comparison of pairs of treatments. The pair comparison is used where there are no common characteristics among the treatments. Sometimes the treatments may have common characteristics in which case the treatments are classified into meaningful groups. Each group will consist of one or more treatments, The aggregate mean of each group is then compared to that of the others. For this purpose the degrees of freedom and sum of squares for treatments are partitioned into components.

The comparison among *t* treatments has the form

$$C_t = w_1T_1 + w_2T_2 + \text{..............} + w_tT_t,$$

where the *w's* are the orthogonal coefficients, *T's* are the totals of the respective treatments, and $\Sigma w_i - = 0$. If C_1 and C_2 are two comparisons such that

$$C_1 = w_{11}\ T_1 + w_{12}T_2 + w_{13}T_3 + \text{............}$$

$$C_2 = w_{21}\ T_1 + w_{22}T_2 + w_{23}T_3 + \text{..........}$$

they are said to be *orthogonal comparisons* if the sum of the products of their coefficients is 0. Symbolically,

$$w_{11}\ w_{21} + w_{12}w_{22} + w_{13}\ w_{23} + \text{..........} = 0.$$

If C_t is any comparison among t treatments, the component of the treatments sum of squares is defined by the expression

$$\frac{C_t^2}{r\left(w_1^2 + w_2^2 + \ldots\ldots + w_t^2\right)} = \frac{C_t^2}{D_t}$$

where, r is the number of replications. Any treatment sum of squares having t - 1 degrees of freedom can be divided into t - 1 orthogonal components. This procedure may be explained with an example.

Consider the fertilizer experiment with 6 treatments-2 organic sources of nitrogen, 3 inorganic sources of nitrogen and 1 non- fertilized control. The comparison of nitrogen sources and control is given by

$$\frac{T_1 + T_2 + T_3 + T_4 + T_5}{5} - T_6$$

where, T's are the totals for the respective treatments. This comparison can be written as

$1 \,.\, T_1 + 1 \,.\, T_2 + 1 \,.\, T_3 + 1 \,.\, T_4 + 1 \,.\, T_5 - 5T_6$

Similarly, the comparison of organic and inorganic sources of nitrogen is given by

$$\frac{T_1 + T_2}{2} - \frac{T_3 + T_4 + T_5}{3}$$

This can be written as

$$3T_1 + 3T_2 - 2T_3 - 2T_4 - 2T_5$$

Observe that the coefficients for one group of treatments is the divisor of the other group. The signs of these coefficients are opposite so that their sum is 0. The above functions are called *linear functions*.

The treatment sum of squares and the degrees of freedom will be partitioned as follows:

Comparison	df	SS
Control *Vs* Fertilizers	1	$\frac{(T_1+T_2+T_3+T_4+T_5-5T_6)^2}{r\left(1^2+1^2+1^2+1^2+1^2+(-5)^2\right)}$
Organic *Vs* inorganic	1	$\frac{(3T_1+3T_2-2T_3-2T_4-2T_5)^2}{r\left[3^2+3^2+(-2)^2+(-2)^2+(-2)^2\right]}$

The remaining degrees of freedom are partitioned as within organic sources and within inorganic sources.

Trends Comparison: It is often desirable to study the nature of response of the experimental units to the varying levels of treatments like rate of fertilizer application, doses of insecticide and dates of harvest. If the treatments are in equal intervals along an ordered scale like 0, 50, 100 and 150 Kg N/ha, the treatment sum of squares may be partitioned into linear, quadratic, cubic, etc., trend components. This method of partitioning provides information about the nature of relationship between treatment levels and treatment means. Trend comparison is applicable only when the treatments are quantitative. The orthogonal coefficients for trend comparisons are taken from tables of orthogonal polynomials.

The numerical examples for above types of comparisons are given under different designs of experiments.

Bibliography

Abbasi, S.A. : *Wetlands of India : Ecology and Threats*, Discovery Publishing House, New Delhi, 1997.

Adkins, R.M. : *Annual Review of Ecology and Systematics*, A. R., London, 1993.

Akhtar, R. : *Disease Ecology and Health*, Rawat Publications, Jaipur, 2001.

Alison, J. : *Spatial Ecological : Economic Analysis for Wetland Management*, Cambridge, London, 2004.

Anthony, J. : *Handbook of Ecological Restoration*, Cambridge, London, 2002.

Antonovics, Janis : *Integrating Ecology and Evolution in a Spatial Context*, Cambridge, London, 2001.

Banerjee, B.N. : *Environmental Pollution and Bhopal Killings*, Gyan Books, New Delhi, 1987.

Barker, G. : *Physiological Plant Ecology*, Cambridge, London, 2002.

Berger, W.H. : *Abrupt Climatic Change*, D. Radiel Dordrecht, Boston, 1990.

Bhargava, Gopal : *Ecological Politics : Different Dimensions*, Gyan Books, New Delhi, 2002.

——— : *Marine Ecosystems*, Gyan Books, New Delhi, 2001.

Bhusan, Bibhuti : *Social Ecology of Forest Resources*, Gyan Books, New Delhi, 2004.

Billihgs, W.D. : *Plant, Man and the Ecosystem,* MacMillan, London, 1971.

Bookchin, M. : *The Philosophy of Social Ecology,* Rawat Publications, Jaipur, 1996.

Chandra, Ramesh and Aneja, Ritu : *Corporate Governance for Sustainable Environment,* Gyan Books, New Delhi, 2004.

Colin, J. : *Issues in Environmental Economics,* Blackwell, London, 2002.

Crawford, T.J. and Hewitt, G.M. : *Genes in Ecology,* Cambridge, London, 1992.

Crosby, A.W. : *Ecological Imperialism : The Biological Expansion of Europe,* Cambridge, London, 2000.

Dakshinamurthi, C. : *Water Resources of India and their Utilization in Agriculture,* Oxford University Press, New Delhi, 1973.

Dass, Sujata K. : *Environment Challenges and Sustainable Future,* Gyan Books, New Delhi, 2003.

Dasseman, R.F. : *Environmental Conservation,* John Wiley, New York, 1976.

Dave, M. : *Urban Ecology and Levels of Development,* Rawat Publications, Jaipur,

David, Khemchand : *Advanced Dictionary of Ecology and Environment,* Dominant Publishers, New Delhi, 2000.

DeVere, Burton : *Ecology of Fish and Wildlife,* Delmar, London, 1996.

Dickinson, L. : *Ecology and Evolution of Cooperative Breeding in Birds,* Cambridge, London, 2004.

Dodson, S. I. : *Ecology,* Oxford, New York, 1998.

Dutta, Ashish : *Biodiversity and Ecosystem Conservation,* Gyan Books, New Delhi, 2001.

Dwivedi, S.D. : *Naturopathy : For Perfect Health,* Gyan Books, New Delhi, 2002.

Eric, R. Pianka : *Evolutionary Ecology,* Harper Collins, London, 1994.

Fritz, L. Knopf : *Ecology and Conservation of Great Plains Vertebrates*, Springer, London, 1997.

Gaan, Narottam : *Environmental Security Concept and Dimensons*, Gyan Books, New Delhi, 2004.

Gary, A. Klee : *The Coastal Environment : Toward Integrated Coastal and Marine Sanctuary Management*, Prentice Hall, London, 1999.

George, A. Maul : *Climatic Change in the Intra : Americas Sea*, Arnold, New Jersey, 1993.

Gerry Closs, Barbare Downes and Andrew, Boulton : *Freshwater Ecology : A Scientific Introduction*, Blackwell, London, 2004.

Gordon, D. : *World Population Problems*, Park Press, Campaign, 1984.

Grafton, Q. : *Economics of the Environment*, Blackwell, London, 2004.

Gupta, K. : *Energy and Environment in India : A Study of Energy Management*, Gyan Books, New Delhi, 2002.

Hourani, G.F. : *Arab Seafaring in the Indian Ocean and Early Medieval Times*, Princeton University Press, Princeton, 1951.

Husain, M. : *Evolution of Geographical Thought*, Rawat, Jaipur, 1995.

Hussen, A. : *Principles of Environmental Economics*, Routledge, New York, 2004.

Hutchings, J. : *The Ecological Consequences of Environmental Heterogeneity*, Cambridge, London, 2000.

Ian, D. Whyte : *Climating Change and Human Society*, Arnold, New Jersey, 1995.

Ian, F. Spellerberg : *Monitoring Ecological Change*, Cambridge, London, 2005.

James, K. Lein : *Integrated Environmental Planning*, Blackwell, London, 2003.

James, M. Bullock : *Dispersal Ecology*, Cambridge, London, 2002.

John, E.FA : *Evolution and Ecology of Macque Societies,* Cambridge, London, 1996.

John, N. Rayner : *Dynamic Climatology : Basis in Mathematics and Physics,* Blackwell, London, 2001.

Joshi, M.V. : *Theories and Approaches of Environmental Economics,* Atlantic Publishers, New Delhi, 2004.

Kaur, Karanjot : *Green Revolution : Ecological Implications,* Dominant Publishers, New Delhi, 1999.

Khanna, D.R. : *Microbial Ecology : A Study of River Ganga,* Discovery Publishing House, New Delhi, 2004.

Kumar, Chattopadhyay : *The Desperate Delta : Social Ecology of Sunderbans,* Gyan Books, New Delhi, 1995.

Kumar, Shashi : *Human Ecology for Globalization : Human Ecology in Action,* Atlantic Publishers, New Delhi, 2003.

——— : *Human Ecology for Globalization : New Dimensions for Economic Growth,* Atlantic Publishers, New Delhi, 2004.

Madan, Mohan : *Ecology and Development,* Rawat Publications, Jaipur, 2000.

Martin, T.E. : *Ecology and Management of Neotropical Migratory Birds,* Oxford, New York, 1995.

Mash, G. P. : *Man and Nature Physical Geography as Modified by Human Action,* Charles Scriber, New York, 1964.

Max, N. : *The Environmental Revolution,* McGraw-Hill, New York, 1970.

Michael, Moss : *Issues and Perspectives in Landscape Ecology,* Cambridge, London, 2005.

Moore, Francis : *Environment and Society,* Dominant Publishers, New Delhi, 2003.

Mukherjee, Roma : *Environmental Management and Awareness Issues,* Sterling Publishers, New Delhi, 2002.

Muthiah, S. : *An Atlas of India,* Oxford University Press, New York, 1992.

Myers, H. : *Ecology and Control of Introduced Plants*, Cambridge, London, 2003.

Nadarajah, M. : *Culture, Gender and Ecology : Beyond Workerism*, Rawat Publications, Jaipur, 1999.

Nigel, J.R. : *Mountains at Risk : Current Issues in Environmental Studies*, Manohar Publishers, New Delhi, 1995.

Nigel, Webb : *Ecological Reviews*, Cambridge, London, 2005.

Overbeck, J. : *Microbial Ecology of Lake Plubsee: Ecological Studies*, Springer, London, 1994.

Pandey, Mahendra : *Environment Pollutants & Women's Health*, Dominant Publishers, New Delhi, 2003.

Paul, A. : *Introduction to Ecology*, John Wiley and Sons, New York, 1973.

Peter, J. Grubb and John, B. Whittaker : *Toward a More Exact Ecology*, Cambridge, London, 1989.

Puri, G. : *Essential Soil Science : A Clear and Concise Introduction to Soil Science*, Blackwell, London, 2002.

Rampinot, M.R. : *Climate: History, Periodicity and Predictability*, Von Nostrand Reinholt, New York, 1987.

Ravi, V.S. : *The Living Adventure : Nature and Biology*, Manohar Publishers, New Delhi, 2001.

Raza, M. : *Development and Ecology*, Rawat Publications, Jaipur, 2000.

Ring, T. Carde and Jocelyn, G. Millar : *Advances in Insect Chemical Ecology*, Cambridge, London, 2004.

Rob, H.G. Jongman : *Ecological Networks and Greenways*, Cambridge, London, 2004.

Sarah, J. Woodin and Mick, Marquiss : *Ecology of Arctic Environments*, Cambridge, London, 1997.

Sheth, P. : *Environmentalism : Politics, Ecology and Development*, Rawat Publications, Jaipur, 1997.

Shukla, S.K. and Srivastava, P.R. : *Ecology and The Environment*, Commonwealth Publishers, New Delhi, 1992.

Sinha, R.K. : *Eco-Politics and Development in India*, Commonwealth Publishers, New Delhi, 1996.

Stamou, G.P. : *Arthropods of Mediaterranean : Type Ecosystems*, Springer, London, 1998.

Thomas, R. : *Man's Impact on the Environment*, McGraw-Hill, New York, 1988.

Thomson, T. : *Introductory Essay to Flora Indica*, MacMillan, London, 1855.

Trivedi, P.R. and Dudarshan, K.N. : *India Ecology and Environment*, Commonwealth Publishers, New Delhi, 1994.

Usher, Michael : *Ecology, Biodiversity, and Conservation*, Cambridge, London, 2005.

Vayda, A.P. G. : *Man in the Pacific Islands: Essays on Geographical Change in the Pacific*, Oxford University Press, New York, 1972.

Venus, W. : *Kanpur City: A Study in Environmental Pollution*, Tara Book Agency, Varanasi, 1982.

Vogt, K.A. : *Ecosystems : Balancing Science with Management*, Springer, London, 1997.

Walter, G.H. : *Insect Pest Management and Ecological Research*, Cambridge, London, 2003.

Wilfried, B. : *Atmospheric Pollution*, McGraw Hill, New York, 1972.

William, W. : *Integrating Landscape Ecology into Natural Resource Management*, Cambridge, London, 2002.

Yadav, P.R. : *Environmental Ecology*, Discovery Publishing House, New Delhi, 2003.

Yeates, M. : *Plant Ecology*, Harper & Row, New York, 1980.

Zelinsky, W. : *A Study of Ecology*, Oxford University Press, New York, 1970.

Zuckerman, Ben : *Human Population and the Environmental Crisis*, Jones & Bart., New Jersey, 1996.

Index

C

D

E

F

N

O

P

Q

R

W

Z

❑❑❑